ATM FOR PUBLIC NETWORKS

ATM for Public Networks

Ronald H. Davis

McGraw-Hill, Inc.
New York · San Francisco · Washington, D.C.
Auckland · Bogotá · Caracas · Lisbon · London
Madrid · Mexico City · Milan · Montreal · New Delhi
San Juan · Singapore · Sydney · Tokyo · Toronto

TK
5105
.35
D29
1999

McGraw-Hill

*A Division of The **McGraw·Hill** Companies*

Copyright © 1999 by Lucent Technologies. All rights reserved. Printed in the United States of America. Except as permitted under the United States Copyright Act of 1976, no part of this publication may be reproduced or distributed in any form or by any means, or stored in a data base or retrieval system, without the prior written permission of the publisher.

1 2 3 4 5 6 7 8 9 0   AGM/AGM   9 0 3 2 1 0 9 8

ISBN 0-07-134476-4

The sponsoring editor for this book was Steve Chapman, the editing supervisor was Sally Glover, and the production supervisor was Pamela Pelton. It was set in Vendome ICG by Kim Sheran of McGraw-Hill's Professional Book Group composition unit, Hightstown, N.J.

Printed and bound by Quebecor Martinsburg.

McGraw-Hill books are available at special quantity discounts to use as premiums and sales promotions, or for use in corporate training programs. For more information, please write to the Director of Special Sales, McGraw-Hill, 11 West 19th Street, New York, NY 10011. Or contact your local bookstore.

 This book is printed on recycled, acid-free paper containing a minimum of 50% recycled, de-inked fiber.

Information contained in this work has been obtained by the McGraw-Hill Companies, Inc. ("McGraw-Hill") from sources believed to be reliable. However, neither McGraw-Hill nor its authors guarantee the accuracy or completeness of any information published herein and neither McGraw-Hill nor its authors shall be responsible for any errors, omissions, or damages arising out of use of this information. This work is published with the understanding that McGraw-Hill and its authors are supplying information, but are not attempting to render engineering or other professional services. If such services are required, the assistance of an appropriate professional should be sought.

This book is dedicated to my parents, Harding and Mary Campbell Davis.

CONTENTS

Preface
Acknowlegements

PART 1   1

Chapter 1   The Evolution of ATM   3

    1. Introduction   4
    1.1 Switching   4
        1.1.1 Circuit Switching   4
        1.1.2 Packet Switching   5
    1.2 Transmission   7
    1.3 The Evolution Communications Services   8
    2. Broadband ISDN   9
    2.1 B-ISDN Services   10
        2.1.1 Interactive Services   10
        2.1.2 Distribution Services 11
    3. Trends Driving the Development of ATM Technology   12
    4. Overview of This Book   14
    References   14

Chapter 2   ATM Technology Concepts   17

    1. Introduction   18
    2. ATM Transport and Switching   18
    2.1 The ATM Switching Concept   20
        2.1.1 Virtual Path Switching   22
        2.1.2 Virtual Channel Switching 23
    3. B-ISDN Network Architecture   23
    3.1 Reference Configuration at the UNI   25
    3.2 ATM Network Elements   26
    3.3 ATM Switching—An Example   27
    3.4 UNI Connection Topologies   29
        3.4.1 Point-to-Point Topologies   29
        3.4.2 Point-to-Multipoint Topologies 30

4.	Quality of Service	32
5.	ATM Signalling: On-Demand Connection Establishment	35
6.	Managing the Network	38
6.1	ATM Network Management Reference Model	39
6.2	Management Information Base	41
6.3	Network Management Activities	42
6.3.1	Configuring Network Elements	42
6.3.2	Permanent Virtual Circuit Establishment	42
6.3.3	Collection of Traffic Performance Statistics	43
6.3.4	OAM Cell Loopback	45
6.3.5	Fault Detection, Reporting, and Recovery	45
References		46

PART II   49

Chapter 3 ATM Communications Protocols   51

1.	Introduction	52
2.	B-ISDN Protocol Reference Model	52
2.1	B-ISDN Protocol Operation	54
3.	The Physical Layer	56
3.1	Physical Media Dependent (PMD) Sublayer	57
3.2	Transmission Convergence (TC) Sublayer	57
3.3	Physical Layer SAP	58
4.	The ATM Layer	58
4.1	ATM Cell Structure	58
4.2	ATM Layer SAP	62
5.	ATM Adaptation Layer	63
5.1	AAL Type 1	65
5.1.1	SAR Sublayer	65
5.1.2	Convergence Sublayer	65
5.1.3	AAL 1 SAP	67
5.2	AAL Type 2	68
5.2.1	Common Part Sublayer	69
5.2.2	Service Specific Convergence Sublayers	72
5.2.3	AAL 2 SAP	78
5.3	AAL Type 3/4	79
5.3.1	SAR Sublayer	80
5.3.2	Convergence Sublayer	81
5.3.3	Connectionless Network Service	82

Contents

5.3.4	AAL 3/4 SAP	84
5.4	AAL Type 5	85
5.4.1	SAR Sublayer	86
5.4.2	Convergence Sublayer	86
5.4.3	AAL 5 SAP	87
5.5	Summary	88
References		89

Chapter 4   ATM Signalling   91

1.	Introduction	92
2.	Signalling Protocol Architecture	92
2.1	Overview of Signalling Architectures	95
2.1.1	Private UNI	95
2.1.2	Private NNI	95
2.1.3	Public UNI	95
2.1.4	NNI	95
2.1.5	B-ICI	96
2.2	ATM Connection Establishment	97
2.2.1	Signalling VCC	97
2.2.2	User Traffic SVC	97
2.3	Signalling ATM Adaptation Layer	98
2.3.1	Service Specific Connection Oriented Protocol	98
2.3.2	Service Specific Coordination Function at the UNI	102
2.3.3	Service Specific Coordination Function at the NNI	103
2.4	Public UNI Signalling	109
2.4.1	Point-to-Multipoint Connection	110
2.4.2	Point-to-Multipoint Connection	112
2.5	Private Network to Network Interface Signalling	115
2.6	Broadband ISDN User Part	116
2.6.1	B-ISUP Specification Model	117
2.6.2	Call Control and Bearer Connection Control	118
2.7	Broadband Inter-Carrier Interface	120
References		120

Chapter 5   Operations Administration, Maintenance, and Provisioning   123

1.	Introduction	124
2.	Operations Administration and Maintenance	126
2.1	OAM Levels and Flows	126
2.1.1	Physical Layer OAM Flows	128

	2.1.2 ATM Layer OAM Flows	128
3.	Provisioning	136
4.	Integrated Local Management Interface	137
References		138

Chapter 6 Traffic Management 141

1.	Introduction	142
2.	Quality of Service Parameters	143
2.1	Sources of Delay	145
2.2	Sources of CDV	146
2.3	Measurement of QoS Parameters	146
2.4	Factors Impacting QoS	147
3.	Traffic Parameters	148
3.1	Peak Cell Rate	149
3.2	Sustainable Cell Rate	149
3.3	Maximum Burst Size	149
3.4	Minimum Cell Rate	149
4.	ATM Service Categories	150
4.1	Available Bit Rate Service	152
	4.1.1 RM Cell Format	153
5.	Usage Parameter Control	155
5.1	Generic Cell Rate Algorithm	157
	5.1.1 Virtual Scheduling Algorithm	157
	5.1.2 Leaky Bucket Algorithm	159
	5.1.3 GCRA Example	160
5.2	Traffic Contract and Conformance Definitions	162
	5.2.1 Conformance Definition for CBR Service	163
	5.2.2 Conformance Definition for rt-VBR Service	163
	5.2.3 Conformance Definition for nrt-VBR Service	164
	5.2.4 Conformance Definition for UBR Service	164
	5.2.5 Conformance Definition for ABR Service	165
5.3	Measurements Associated With UPC/NPC	166
6.	Connection Admission Control	166
7.	Network Resource Management	167
8.	Traffic Shaping	167
9.	Selective Cell Discard	168
10.	Frame Discard	168
11.	Guaranteed Frame Rate Service	168
References		170

Contents

PART III   171

Chapter 7   ATM Service Interworking   173

    1. Background   174
    2. Circuit Emulation Service   176
    2.1 Structured CES   179
        2.1.1 Basic CES   181
        2.1.2 CES with CAS   181
        2.1.3 ISDN CES   186
    2.2 Unstructured CES   186
    2.3 Variable Bit Rate CES   187
    2.4 ATM Trunking   188
        2.4.1 Channel Associated Signalling   188
        2.4.2 Narrowband ISDN Signalling   191
    2.5 CES Service Definition   194
    3. Frame Relay Service   195
    3.1 Frame Structure   195
    3.2 Frame Relay Service Specific Convergence Sublayer   198
        3.2.1 Discard Eligibility (Frame) To Cell Loss Priority (Cell) Mapping   198
        3.2.2 Forward Explicit Congestion Notification (Frame) To Explicit Forward Congestion Indication (Cell) Mapping   199
        3.2.3 Backward Explicit Congestion Notification (Frame) To Explicit Forward Congestion Indication (Cell) Mapping   200
    3.3 Frame Relay and ATM Interworking: Network View and Service View   201
        3.3.1 Implementation Agreement FRF.5   201
        3.3.2 Implementation Agreement FRF.8   203
    3.4 FRS Service Definition   204
    4. ATM Frame Based User to Network Interface   206
    5. Switched Multi-megabit Data Service   207
    5.1 SIP L3_PDU Encapsulation   208
    5.2 SMDS to ATM Mapping   210
    5.3 ATM Adaptation Layer   210
    5.4 SMDS Service Definition   211
    References   212

Chapter 8   IP Over ATM   215

    1. Addressing within ATM Networks   216

2.	Routing Within ATM Networks	218
2.1	Discovery of Neighbors and Link Status	220
2.2	Synchronization of Topology Databases	221
2.3	Flooding of PTSEs	221
2.4	Election of Peer Group Leader	222
2.5	Summarization of Topology State Information	223
2.6	Construction of the Routing Hierarchy	224
2.7	Route Selection	225
3.	Classical IP over ATM	226
3.1	Communication between Stations in the Same LIS	227
3.2	Communication between Station in Different LIS'	229
4.	Next Hop Resolution Protocol	230
5.	LAN Emulation	235
5.1	LANE Protocol Summary	236
5.2	LANE Example	238
6.	Multicast IP Over ATM Networks	241
6.1	Review of IP Multicasting	241
6.2	ATM Multicast IP Configuration	241
7.	Multiprotocol over ATM	244
7.1	MPOA Data Flow	247
7.2	MPOA Communication Scenarios	248
7.2.1	Case 5: MPOA Host (to MPOA Host, different ELANs	249
7.2.2	Case 6: LAN Host to MPOA Host, different ELANs	249
7.2.3	Case 7: MPOA Host to LAN Host, different ELANs	249
7.2.4	Case 8: LAN Host to LAN Host via Edge Devices in different ELANs	250
8.	Multiprotocol Encapsulation over ATM	250
8.1	LLC/SNAP Encapsulation	251
8.2	VC Based Multiplexing	251
9.	Interaction between ATM Routing and IP Routing	252
	References	252

Chapter 9 TCP Over ATM 255

1.	Introduction	256
2.	TCP Features	256
2.1	Sliding Window Flow Control	256
2.2	TCP Window Size	257
2.3	TCP Segment Size	258
2.4	TCP Behavior	259

Contents

	2.4.1	Slow Start	260
	2.4.2	Congestion Avoidance	260
	2.4.3	Fast Retransmit	261
	2.4.4	TCP Performance	262
3.	TCP Performance Over ATM Networks		262
3.1	QoS and TCP Performance		263
3.2	Cell Dropping Strategies and TCP Performance		263
	3.2.1	Plain ATM	264
	3.2.2	Partial Packet Discard	264
	3.2.3	Early Packet Discard	265
	3.2.4	Limitations of PPD and EPD	266
3.3	TCP Over ATM Performance: Empirical Results		266
3.4	Simulation of TCP over ATM Performance		267
	3.4.1	TCP over ATM in a LAN Environment	267
	3.4.2	TCP over ATM in a Short Haul Environment	269
4.	Conclusion		274
References			275

Chapter 10 ATM and the Internet 277

1.	Introduction	278
2.	Evolution of the Internet	278
3.	Internet Standards	280
4.	Internet Traffic Patterns	280
5.	The Integrated Services Internet	282
5.1	Packet Scheduler	283
	5.1.1 Packet Discard	285
5.2	Packet Classification	288
5.3	Admission Control	288
6.	Resource Reservation Protocol	289
6.1	Path Message	291
	6.1.1 Sender Tspec	292
	6.1.2 Adspec	293
6.2	Resv Message	295
	6.2.1 Flowspec	295
6.3	Initiation of an RSVP Session	297
7.	Integrated Internet Services Over ATM	297
7.1	VC Management	298
	7.1.1 PVC Versus SVC	298
	7.1.2 ATM Multipoint and RSVP Heterogeneous Branch Points	299

	7.1.3	Dynamic "Soft State" Versus Static QoS	301
	7.1.4	Signalling Traffic	302
	7.1.5	Synchronizing ATM VC and RSVP States	303
7.2	QoS Management	303
	7.2.1	Guaranteed Service	303
	7.2.2	Controlled Load Service	305
	7.2.3	Best Effort Service	305
	7.2.4	Traffic Tagging	305
8.	vBNS	306
9.	ATM Deployment At Internet Network Access Points	308
10.	IP/ATM and IP/SONET	311
10.1	Protocol Efficiencies	311
	10.1.1	SONET Overhead	311
	10.1.2	IP/PPP/SONET Overhead	311
	10.1.3	IP/ATM/SONET Overhead	312
	10.1.4	Comparative Efficiencies IP/PPP/SONET versus IP/ATM	313
	10.1.5	Empirical Results	313
10.2	IP/SONET versus IP/ATM: Which Is Better?	314
References	315

Chapter 11   Voice Over ATM   319

1.	Introduction	320
2.	Voice over ATM to the Desktop	320
3.	Recommendation H.323—Voice over the Internet	322
3.1	Description of H.323	324
	3.1.1	Call Setup	325
	3.1.2	Initial Communication and Capability Exchange	327
	3.1.3	Establishment of Audiovisual Communication	327
	3.1.4	Call Termination	330
4.	H.323 Over ATM	330
4.1	Protocol Efficiency	333
4.2	Interworking H.323 and Non-H.323 Terminals	335
References	335

PART IV   **337**

Chapter 12   Future Directions for ATM   339

| 1. | Introduction | 340 |

Contents

1.1	Managed Bandwidth Versus Infinite Bandwidth	340
2.	Emerging Applications	342
2.1	Asymmetric Digital Subscriber Line	342
2.2	Digital Television	343
2.3	Wireless ATM	344
2.3.1	BSC Interconnection	344
2.3.2	ATM Interworking with PCS	345
2.3.3	ATM over LMDS	345
3.	Future Challenges	346
	Acronyms	349
	Glossary	355
	Index	375

Preface

Objectives of This Book

The primary objective of this book is to present a discussion of the role of ATM technology in public networks. The topics presented in this book, however, touch upon a number of issues that are relevant in both public and private networks. The discussions presented here center upon three major subject areas: voice over ATM, data over ATM, and interworking of existing services over ATM networks.

The integration of voice and data over a common network infrastructure is a driving factor toward the evolution of communication networks. While data traffic is the much faster-growing component and is expected to dominate network traffic volume, the basic role of the communications network in providing a medium for communication between people will not change. Thus, the performance of a communications network in providing basic voice telecommunications service will remain critical. Voice and data have different traffic characteristics and present very different issues when attempting to carry both over a common communications network infrastructure. One of the objectives of this book is to present the issues involved in carrying voice traffic and data traffic over ATM networks.

Within the public network environment are dedicated networks that support current voice and data service offerings. An important issue in the adoption of ATM technology in the public networks is the ability of ATM to interwork with these existing networks and the services provided over those networks. In this book we discuss two approaches to interworking of existing services over ATM networks: network interworking (or tunneling), in which traffic from other networks is carried transparently across an ATM network; and service interworking, in which the services carried by other networks are translated into equivalent ATM services for transport across an ATM network.

What this book does not purport to do, however, is to give the reader "the solution," or to prescribe a course of action for the reader to follow. Rather, the objective of this book is to provide the reader with a conceptual framework for thinking about ATM technology: how it could be used, where it should be used, and to position ATM versus comparable technologies. Thus, the intention of the author is to present information about ATM and relevant alternative technologies in as objective a manner as possible without taking a position on whether ATM is an inherently "better" technology.

Intended Audience

ATM for Public Networks presents topics that will be of interest to data communications managers, network design engineers, and those who have a general interest in gaining a broader understanding of the technologies that will influence the rapidly evolving communications industry.

This book will be of interest to Internet Service Providers who want to understand how ATM technology may be applied to Internet backbones, as well as to Incumbent and Competitive Local Exchange Carriers (ILEC/CLEC) who are considering migration of existing services over dedicated networks to a unified network infrastructure that can also serve as a platform for new and expanded service offerings.

Software and hardware developers will be interested in the detailed protocol discussions that are presented in this book. There is also a comprehensive list of references that will be of great use to implementors of protocol stacks.

The technology comparisons, which describe some of the advantages and disadvantages of ATM versus alternative technologies, will be of interest to managers, network planners, and potential customers of ATM services and products. This book also explains the principles and operation of important emerging technologies that will be of interest even to those who don't have specific ATM-centric interests.

It is assumed that the reader has prior knowledge of the fundamentals of telecommunications principles and data networks.

Organization of This Book

In the first two chapters of this book, we present an introduction to ATM and an overview of the key concepts associated with ATM technology. Chapter 1 describes the evolution of switching technology from circuit switching to ATM. This chapter also describes the trends and services that spurred the development of ATM technology.

Preface

Chapter 2 presents a broad overview of the key technology concepts that define ATM. This chapter provides an introduction to ATM network architecture, switching concepts, quality of service, ATM signalling, and network management. In chapter 2 we set the stage for chapters 3-6, which discuss these topics in further detail.

Chapter 3 begins the detailed discussion of ATM protocols with an introduction to the ATM Protocol Reference Model (PRM) and the User Plane, Control Plane, and Management Plane. Chapter 3 then goes on to focus on a description of the User Plane protocols. The User Plane represents the set of functions that provide for the transfer of information between communicating endpoints. This chapter describes the ATM layer and ATM Adaptation Layer (AAL) protocols. The AAL discussion explains the AAL types, which are defined at the time of this writing as Type 1, Type 2, Type 3/4, and Type 5.

Chapter 4 presents a discussion of the PRM Control Plane, which provides procedures for ATM signalling. This chapter describes the Signalling AAL protocol stacks, or SAAL, for each type of User Network Interface (UNI) and Network Node Interface (NNI) that is present in the ATM network architecture. Also presented in this chapter is a discussion of how ATM signalling procedures enable services for point-to-point and point-to-multipoint connections.

Chapter 5 describes the Management Plane which supports network management capabilities. In this chapter we describe protocols for performing Operations, Administration, Maintenance, and Provisioning (OAM&P) in ATM networks. Included is a discussion of the methods and capabilities for performance monitoring of ATM networks. The discussion of OAM&P in this chapter is presented in relation to the Telecommunications Management Network (TMN) model; however, Chapter 5 also describes the SNMP-based Integrated Local Management Interface (ILMI) protocol.

Chapter 6 discusses the portion of the Management Plane that provides traffic management capabilities. This chapter describes Quality of Service (QoS) and the traffic management functions performed within ATM networks. Chapter 6 also describes the service classes defined for ATM virtual connections: Constant Bit Rate (CBR), Real-Time Variable Bit Rate (rt-VBR), Non-Real-Time Variable Bit Rate (nrt-VBR), Unspecified Bit Rate (UBR), and Available Bit Rate (ABR), relating each service class to a corresponding QoS specification. A grasp of the concepts presented in this chapter provides the basis for understanding how QoS is provided to traffic carried over ATM networks. In this chapter we also describe the Guaranteed Frame Rate (GFR) Service.

Using the discussions of the previous chapters as a foundation, Chapter 7 begins the discussion of how existing services may be interworked over a unified ATM network infrastructure. The chapter describes the interworking of circuit-switched services, Frame Relay, and SMDS over ATM networks. We also discuss an important, frame-based, native ATM subscriber interface: the Frame-Based UNI.

The next two chapters discuss the use of ATM to carry TCP/IP traffic. However, unlike the typical treatment of this topic in other publications, we discuss the interactions between TCP and IP with ATM as separate discussions in order that the reader may clearly see how TCP and IP interact differently with ATM. Chapter 8 is concerned with IP traffic over ATM networks. The chapter begins with a discussion of ATM address formats and proceeds to a discussion of routing within private ATM networks. Here we describe the Private Network Node Interface (PNNI) protocol. Next we describe IP over ATM methods that includes discussion of Classical IP over ATM, Next Hop Resolution Protocol (NHRP), LAN Emulation (LANE) and Multiprotoocol over ATM (MPOA). There is also a discussion of how multicast IP traffic is carried over ATM networks. The chapter includes a discussion of the interactions between ATM routing and IP routing. In Chapter 8 we generalize this discussion by explaining how Network Layer protocols other than IP may be encapsulated over ATM networks using a technique called LLC/SNAP encapsulation.

Chapter 9 discusses the interactions between TCP and ATM. We begin the chapter with a review of TCP with an emphasis on the bandwidth control features of TCP. The discussion then proceeds to describing TCP performance over ATM networks. This description focuses upon the interaction between TCP behavior and the QoS mechanisms of ATM. This presentation also describes traffic management methods that increase the efficiency with which ATM networks carry TCP traffic: Partial Packet Discard (PPD) and Early Packet Discard (EPD). Included are very surprising empirical results from studies of TCP performance over ATM networks. Chapter 9 concludes with a review of literature describing simulations of TCP over ATM networks which suggest some of the impacts of traffic management strategies on ATM switch design.

Chapter 10 discusses how ATM may be used in Internet backbone networks. The chapter begins with a brief history of the evolution of the Internet, which leads to a description of the Integrated Services Internet. The Integrated Internet Service (IIS) model is a proposal for a next-generation network that will have capabilities that enable the Inter-

Preface

net to carry traditional data traffic and to provide real-time services. Our IIS discussion includes descriptions of methods for traffic management, such as Weighted Fair Queueing (WFQ), Random Early Drop (RED), and Weighted RED (WRED). This chapter also includes detailed discussion of the Resource Reservation Protocol (RSVP). Next, we describe how the IIS model may be deployed over ATM networks. This discussion focuses on the interworking of IIS with ATM.

Chapter 10 then presents two case studies of ATM deployment in the Internet. The first case study describes the deployment of ATM in an Internet backbone, while the second describes the deployment of ATM at an Internet Network Access Point.

An important topic of discussion in consideration of ATM deployment in the Internet involves a comparison between IP over ATM, versus IP/Packet over SONET. In Chapter 10 we compare the two alternatives and present the results of an empirical study of the comparative efficiencies of the two approaches.

Chapter 11 describes the use of ATM networks to carry voice traffic. This chapter discusses the deployment of voice over ATM at the desktop and presents a description of the interworking that enables end-to-end voice communications between ATM Networks and traditional Public Switched Telephone Networks (PSTN).

An important proposal for carrying voice over the Internet is described in Recommendation H.323 from the International Telecommunication Union (ITU). In Chapter 11 we describe the architecture and operation of H.323 networks, and we describe how H.323 may be deployed over an ATM infrastructure. At the end of Chapter 11 we present a comparison of the efficiency of the different methods of carrying voice described in this chapter.

Chapter 12 presents a retrospective summary of the issues discussed in the book. In this chapter we also describe emerging applications for ATM, including a discussion of "last mile" deployment of ATM to customer premises using ADSL technology.

Acknowledgments

The creation of a book requires the devotion of vast amounts of time and energy. While the name of the author appears on the book, there are also the names of the many people who contributed their time, knowledge, and experience, and I would like to take this occasion to thank them. In particular I would like to thank Bob Rodgers from Chrysler; Thomas Archuleta, Steve Gossage, and John Naegle from Sandia National Laboratories; Andy Schmidt from Ameritech Advanced Data Services; and Dr. William Fisher, Mike Marcinkevicz, Don Pardoe, and Dave Reese of the California State University System. In addition, I would like to thank the participants of the comp.dcom.cell-relay Internet newsgroup for their questions, answers, and general discussions that contributed to the content of this book.

I would also like to give special thanks to Terry MacGregor of Lucent Technologies for his comments, suggestions, and overall motivational support that helped keep me going during the writing of this book; Tom Lyons, and Bob Dianda, also of Lucent Technologies, who provided valuable subject matter expertise in the areas of voice over ATM, and ATM signalling; Richard Sanchez of Cisco Systems, who was there at the beginning of this project and helped me to get started; and Gary Kessler of Hill Associates, who reviewed the entire book and provided many helpful comments and suggestions.

For strictly personal reasons I would like to thank Professor Gerald Sussman of MIT, and Victor Ransom, now retired from Bell Laboratories, each who gave me an opportunity during my early years as a student at MIT that, when viewed in retrospect, proved to be critical events that shaped the course of my life and career. I later came to think that if I ever got an opportunity to thank them publicly that I should do so, and this work, being my first published book, has created just such an opportunity.

CHAPTER 1

The Evolution of ATM

1. Introduction

Asynchronous Transfer Mode (ATM) is a packet-oriented transfer mode that uses an asynchronous time division multiplexing technique. A *transfer mode* refers to a set of methods covering transmission, multiplexing, and switching in a telecommunications environment. As such, ATM is the culmination of developments in switching and transmission technologies. The result, ATM, is a technology that has, in turn, enabled telephony, video distribution, and data services to be carried over a single communications network.

In order to understand ATM, we will first discuss the switching and transmission technologies that made it possible. Next we will discuss the services that are supported by ATM.

1.1 Switching

Switching refers to the process of connecting appropriate transmission facilities to form a desired communication path between two communicating endpoints. Switching technologies falls into two major categories: circuit switching and packet switching [1][2].

1.1.1 Circuit Switching

In circuit switching, resources in a network are dedicated to the parties involved in a transfer for the duration of the communication. For this reason, circuit-switched connections are referred to as *physical* connections. In a Time Division Multiplexing (TDM) network, the bandwidth available on the network is divided into fundamental units called *timeslots*. Groups of timeslots are grouped into *frames*, as shown in Figure 1-1, for efficient transport across the network. Frames, and the timeslots within them, are constantly transmitted across the network at fixed time intervals.

In a circuit-switched network, specific timeslots within a frame are allocated for exclusive use on a channel between two communicating endpoints. Consequently, the only information needed within the network to transfer information over a specific channel is identification of the timeslot number (or numbers) within a frame.

Circuit-switched networks are very efficient at delivering information with little end-to-end delay, and with no variation in the amount of

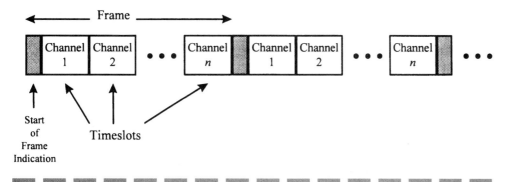

Figure 1-1. Circuit-switched transfer.

delay for delivery of timeslot information. This makes circuit switching well suited for voice telephony applications and for other services that send information at a constant bit rate (CBR). For applications such as data communications, however, traffic is generated from highly bursty, variable bit rate (VBR) sources. In these cases, circuit switching is wasteful of network bandwidth as many timeslots may go unused. From the network operations perspective, this is undesirable since the total number of connections that may be supported is limited by the number of timeslots available in the frame. Thus, when a timeslot limit is reached, the network operator must deploy additional network resources to serve more customers. From the subscriber perspective, circuit-switching services require that the customer contract for service be at the maximum bit rate, even if the capacity is underutilized.

Circuit switching is a *synchronous* (as opposed to asynchronous) method of information transfer because user information is transferred across the network in specific timeslots, and only on those timeslots.

1.1.2 Packet Switching

An alternative to circuit switching that is more appropriate for VBR traffic is Packet Switching. Packet switching does not reserve network resources to a single connection, but rather, allows timeslots to be used on an *as-needed* basis, as shown in Figure 1-2. Since the destination of a given *packet* of information is no longer indicated by a specific timeslot assignment, the packet itself must contain header information that

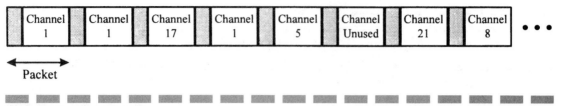

Figure 1-2. Packet transfer.

identifies the destination. Such header information is carried in addition to the payload information.

Packet-switching services may be either of the following:

- Connection-oriented, in which a connection between the two communicating endpoints is established prior to any information transfer. In a Connection-Oriented Network Service (CONS), all packets follow the same path through the network. This also guarantees that all packets arrive at the receiving end in the same order as sent from the transmitting end. Since network resources are not committed to a particular connection as they are in circuit switching, these connections are referred to as *virtual* connections.

- Connectionless, which allows information exchange between communicating endpoints without establishment of a connection. Connectionless Network Service (CLNS) allows information exchange without the delay required while setting up a connection. In addition, packets are not constrained to follow the same route between communicating endpoints; each packet may be routed independently. As a consequence, packets may arrive at the receiving end out of sequence.

In packet-switched networks, resources may be more efficiently utilized by engineering the network such that the available bandwidth on the network is less than the aggregate peak bit rates available to the subscribers. This is based upon the assumption that not all subscribers will be sending information at the peak bit rate at the same time, and that periods of time during which a significant number of subscribers are sending peak rate bursts will be sufficiently short so as not to exceed the capacity of the network to absorb them. Packet-switched networks use a process called statistical multiplexing to allocate network resources to subscribers based upon the premise that traffic from subscribers is statistically distributed with a mean bit rate that is less than the peak rate. By engineering the network according to statistical distributions of anticipated subscriber traffic, network operators are able to utilize network resources more efficiently than would be the case with circuit-switched networks. Since packet-switched networks allocate resources in response to the arrival of subscriber information, which may occur at any time, this method of information transfer is referred to as being *asynchronous*.

1.2 Transmission

Transmission is the process of sending information from one point to another in a communications network over some medium. This includes connection of one communications switch to another (referred to as a *trunk*) or between a subscriber and a switch (called a *loop*).

Digital, or T-carrier transmission systems were first designed in the 1960s. These systems provided economical transport of information between switching offices by multiplexing and demultiplexing traffic from a large number of channels onto a common transmission medium. T-carrier transmission systems operated at different bit rates are referred to as Digital Signal (DS) levels. This led to the development of a hierarchy of DSn rates in what is known as the Plesiochronous Digital Hierarchy (PDH). The PDH for North America is shown in Table 1-1 [3]. Rates range from DS0, the channel rate for voice telephony (64 kbps), to DS4NA (139.264 Mbps). While twisted-wire pair provides a transmission medium sufficient for DS0 rate traffic, the highest rate, DS4NA, is carried exclusively over coaxial cable.

TABLE 1-1

Plesiochronous Digital Hierarchy (North America)

DS Level	Bit Rate (Mbps)	Equivalent Voice Channels
DS0	0.064	1
DS1	1.544	24
DS1C	3.152	48
DS2	6.312	96
DS3	44.736	672
DS4NA	139.264	2016

Because it is plesiochronous, individual DSn channels were not synchronous with respect to higher (or lower) signal rates in the hierarchy. Adding a DS1 channel to a DS3 channel required demultiplexing the DS3 down to constituent DS1s, resynchronizing all DS1 channels, and then remultiplexing the DS1s back into a DS3. Thus, adding and removing channels requires network operators to invest in equipment to multiplex and demultiplex signals at various levels in the PDH.

Another disadvantage of PDH is that three different hierarchies were developed in North America, Europe, and Japan, resulting in costly interfaces among signals from different hierarchies.

By the 1980s, advances in transmission equipment and the development of fiber-optic cable allowed for higher transmission rates (along with lower bit error rates). This led to the development of a new digital hierarchy. In North America, this new hierarchy was called the Synchronous Optical NETwork, or SONET. The related international standard was called the Synchronous Digital Hierarchy (SDH). The SONET hierarchy is shown in Table 1-2 [4]. While SONET and SDH use different terminologies, the bit rates are compatible. SONET defines the Synchronous Transfer Signal (STS) as the basic building block defining rates from STS-1 (51.84 Mbps), to STS-192 (nearly 10 Gbps).

1.3 The Evolution in Communications Services

The different methods of routing used in circuit-switched and packet-switched networks resulted in separate, dedicated-purpose networks to handle each type of traffic. Circuit-switched networks were built upon networks using TDM switches, while packet networks used

TABLE 1-2.

SONET Hierarchy

STS Level	Bit Rate (Mbps)	Equivalent Voice Channels
1	51.84	672
3	155.52	2016
12	622.08	8064
48	2488.32	32256
192	9953.28	129024

packet switches. Consequently, the subscribing customer had to also maintain different Customer Premises Equipment (CPE) to interface to each network.

Within data networks, subscribers to early packet-switched services, such as X.25, were limited to rates no higher than DS0. These protocols were the product of the transmission systems that existed in the 1970s. The bit error rates of transmission facilities were such that extensive error checking was required to be performed at each switching node encountered as the packet passed through the network. As later transmission systems provided improved bit error rate performance, newer, faster packet-switched services, which performed fewer error checks, became viable.

One such service was Frame Relay, which sends information in variable sized packets known as *frames*. The reduced amount of error checking allowed frames to be switched at a faster rate at network nodes, which in turn resulted in reduced end-to-end delay, or *latency*. The transport of variable size frames, however, has the disadvantage of creating variations in the end-to-end delay encountered by individual packets as they are transported across the network. This is acceptable for data applications that are not sensitive to variations in packet transit time from the transmitting end to the receiving end. This is not acceptable for isochronous applications such as voice, which depend upon a predictable and constant end-to-end delay.

This led to the development of yet another packet-switched service, Cell Relay. In Cell Relay, all packets (referred to as a *cell*) are of fixed size. This reduces the source of variation in end-to-end transit delay as discussed in Frame Relay. Small cell sizes are desirable for isochronous applications since these cells can be switched rapidly at network nodes. In data applications, however, large cell sizes are more efficient because they offer higher ratios of user information octets to overhead (header information) octets. This means that more of the total transmission bandwidth is available for the transfer of user information.

In ATM, a compromise was sought that would meet the needs of both voice and data applications. The result was a cell size of 53 octets [5]. In addition, ATM defined a series of service classes for the different types of traffic that were to be supported by an ATM-based service. These included support for voice, video, and data in a single network.

2. Broadband ISDN

The mid-1980s ushered in developments leading to a new network service that would merge what had previously been separate voice and data networks. This was called the Integrated Services Digital Network (ISDN). But while ISDN provided subscribers with a single, digital access channel for both voice and data traffic, the underlying communications network continued to maintain an infrastructure of separate circuit-switched and packet-switched networks.

Subscribers of ISDN services were also limited to bit rates no higher than DS1. However, as developments in transmission led to higher bit rates, there was also the desire to deliver information to subscribers at rates higher than DS1. This extension of higher bit rate services to subscribers was called Broadband ISDN (B-ISDN).

Realizing B-ISDN, however, meant that corresponding developments in switching technology were needed. Thus, ATM evolved into a solution to the switching problem by providing an underlying structure for the fast switching of information packets. ATM is a specific packet-oriented transfer mode that uses asynchronous time-division multiplexing techniques [6]. Combined with the definitions of service classes, ATM allowed B-ISDN to meet its two primary objectives:

1. *Delivery* of high bit rate services to subscribers.
2. *Integration* of voice, video, and data over a single network infrastructure.

Because of the close connection between ATM and B-ISDN, the terms are often used interchangeably, with the term ATM being more commonly used.

2.1 B-ISDN Services

The early 1990s view of B-ISDN was that it was intended to provide services that fell into two categories: interactive services and distribution services. In turn, these categories were further subdivided into subcategories [7].

2.1.1 Interactive Services

2.1.1.1 Conversational Services These are services that support bidirectional communication with real-time end-to-end information exchange. Examples of conversational services would be videotelephony, videoconferencing, and high-speed transactional data. The applications for such services include distance learning involving live instructors, "shop at home," building monitoring, traffic monitoring, and real-time control. In addition, there are numerous applications involving collaboration among individuals. This includes, CAD/CAM, motion picture editing, and "telemedicine," which would allow physicians to consult on matters of patient diagnosis, etc. There are a number of entertainment applications, such as interactive game playing, which would also use conversational services.

2.1.1.2 Message Services These services would not be real time, but would allow a sender to deposit a message to a "mailbox," where it could be retrieved any time afterwards by the intended receiver. In addition to traditional telephony voice mail services, the B-ISDN service would allow messages to consist of motion pictures, high-resolution images, and high-quality audio information. Applications of these services would include "video mailbox" services, or document mailbox services, which allow documents to consist of mixed media types (voice, video, fax, data).

2.1.1.3 Retrieval Services This service is a variation of the message service in which information is effectively held in a "public" mailbox for retrieval by all users who have been granted the authorization to retrieve information. An example of a retrieval service would be a publicly available archive of video, and/or audio information. Applications of retrieval services could include distance learning involving stored or prerecorded information, news retrieval, and remote downloading of software.

2.1.2 Distribution Services

2.1.2.1 Distribution Services without User Presentation Control
These include broadcast services in which authorized users would be able to retrieve information from centralized broadcast servers. Users, however, would be unable to control the sequence in which information is retrieved from the server, or the time at which the string of information is started (e.g., the start time of a movie or television program). Television and radio broadcast are examples of services in which the presentation is not under user control, but rather based upon a schedule established by the broadcast operator. Broadcast of television programs using existing standards (NTSC, PAL, SECAM), as well as high-definition TV, are applications of this service subclass. Premium (i.e., pay) TV services, which charge

a fee for viewing a specific program or for viewing a specific channel for a block of time, are also potential applications.

2.1.2.2 Distribution services with User Presentation Control are similar in concept to retrieval services. The key difference is that the distribution service provides a sequence of information with cyclical repetition. Unlike distribution services without user presentation control, this service would allow each individual user to retrieve information starting from any point in the information sequence. A service that displays movies on-demand of individual subscribers would be an example of this subclass of service. Self-paced distance learning is also an application for distribution services with user presentation control.

3. Trends Driving the Development of ATM Technology

While Broadband ISDN is seen as a service for both business and residential subscribers, ATM technology is being driven primarily by the needs of business communications. The market for ATM-based products and services for business communications may be divided into three segments:

1. **Public Network Infrastructure.** This is the core network deployed by operators to support telecommunications services such as public telephony, data services, videoconferencing, etc. Equipment in this category includes central telephone office switches, remote switches, access devices, multiplexers, and line interfaces.

2. **Local Area Networks.** This segment may be defined as the equipment used to interconnect personal computers and workstations, as well as other computing resources over relatively short distances (within a building, or within a campus of buildings). The equipment in this category includes all classes of computers, network interface cards, hubs, switches, bridges, and routers.

3. **Wide Area Networks.** These are used to interconnect remote sites over lines that are typically leased by a business or enterprise from a public network operator.

More bandwidth is being used within public and private networks. Bandwidth expansion is being led by a growth in data traffic due to the expansion of the Internet, as well in as the private network equivalent, corporate intranets. Interconnection of geographically dispersed local area networks is another bandwidth driver.

The trend towards networked computing, in which workstations and personal computers retrieve data from centralized "servers," is increasing bandwidth needs within campus networks, and in some cases, across wide area networks. In many such networks, the objective is to create a "virtual" disk drive, in which data accesses from servers occur as quickly as though the data were being retrieved from a disk drive directly attached to the workstation. Furthermore, the prospect of a network computer, which retrieves both data and applications software from servers, will result in even greater bandwidth needs on networks.

As increasing amounts of computer processing power are becoming available to individual users, larger, more data-hungry applications are being created. These applications, such as CAD/CAM are using data files that are approaching gigabyte size.

In many corporate environments network expansion plans are being developed to anticipate growth in the amount of video traffic—for videoconferences and information distribution.

These trends, however, merely speak to higher bandwidth—in which case the solution would ostensibly be as simple as increasing the amount of bandwidth by extending bigger bandwidth pipes to the subscriber (e.g., replacing a DS1 line with a DS3, or OC3). There is a limit, in some cases, to the degree to which bandwidth alone can be looked to as "the answer." In applications where the great majority of the total processing time is due to computations performed at the workstation, such as compute-intensive applications like CAD/CAM, additional bandwidth is not the solution and yields only modest increases in processing throughput.

When bandwidth is the root problem, however, ATM supports a Quality of Service (QoS) concept, which is a mechanism for allocating resources based upon the needs of the specific application. This differs from earlier packet-switched services which merely allocated network resources on a first-come, first-served basis. ATM goes a step further by assigning a priority (or QoS) to network traffic and allocating resources to the highest priority traffic first—a factor that becomes important as applications require increasingly more bandwidth on networks.

4. Overview of This Book

In the first part of this book we present an introduction to ATM with this chapter, presenting a context and motivation for ATM. In Chapter 2, we present an overview of concepts associated with ATM technology.

The second part of this book consists of four chapters that describe ATM protocols. Beginning with Chapter 3, we present an introduction to the ATM protocol reference model (PRM) and describe user plane ATM and ATM Adaptation Layer (AAL) protocols. Chapter 4 focuses on the control plane portion of the ATM PRM, signalling. Chapters 5 and 6 describe protocols related to the management plane portion of the PRM: Chapter 5 discusses ATM network operations administration, maintenance, and provisioning, with the subject of Chapter 6 being traffic management.

The third part of the book begins in Chapter 7, which describes the capabilities of ATM to provide a multiservice infrastructure for the support of both ATM services and for existing communications services. Chapter 8 describes IP over ATM protocols, Chapter 9 discusses the interactions between TCP and ATM, and Chapter 10 describes the implications of ATM deployment in the Internet. Chapter 11 discusses voice over ATM and the interworking of ATM and H.323 networks.

The final part of the book presents a summary and discussion of future applications for ATM.

References

1. Rey, R.F. ed., *Engineering and Operations in the Bell System, 2d Ed.*, Bell Laboratories, Inc., 1983.
2. Stallings, William *Data and Computer Communications, 3d Ed.*, Macmillan Publishing Co., 1991.
3. Digital Hierarchy—Formats Specifications, American National Standards Institute, ANSI T1.107, July 1995.
4. Synchronous Optical Network (SONET)—Basic description including Multiplex Structure, Rates, and Formats, American National Standards Institute, ANSI T1.105, October 1995.

5. De Prycker, Martin, *Asynchronous Transfer Mode, 2d Ed.,* Ellis Harwood Ltd., 1993.
6. B-ISDN Asynchronous Transfer Mode Functional Characteristics, International Telecommunication Union, Recommendation I.150, November 1995.
7. B-ISDN Service Aspects, International Telecommunication Union, Recommendation I.211, March 1993.

CHAPTER 2

ATM Technology Concepts

1. Introduction

The chapter presents an overview of key concepts that define ATM technology. The intention of this chapter is to give the reader an understanding of the structure and mechanisms employed in ATM communications. Specific topics discussed are the following:

- ATM Transport and Switching
- B-ISDN Network Architecture
- Quality of Service
- ATM Signalling
- Operations Administration, Maintenance, and Provisioning

Further detail on the topics discussed in this chapter can be found in the following chapters.

Standards that define ATM networks are established by two different organizations:

- International Telecommunication Union (ITU)—Formerly known as the CCITT, this organization issues international standards (called "recommendations") for telecommunications.
- ATM Forum—An industry consortium that is focused on ATM. The activities of this organization included setting of technical standards for ATM systems and facilitating the market adoption of ATM technology.

The two organizations frequently use different terminology to describe the same concepts. In this chapter, and throughout this book, we will attempt to distinguish ITU and ATM Forum terminology and concepts where appropriate.

2. ATM Transport and Switching

An ATM transport network is structured into two layers: an ATM Layer, which involves the switching aspects of the network, and a Physical Layer, which involves the transmission aspects. ATM Layer transport functions are independent of the Physical Layer. The ATM Layer is fur-

Chapter 2: ATM Technology Concepts

ther subdivided into two levels that represent the switching levels supported within the network [1]:

- Virtual Channel (VC)—a generic term used to describe a capability for the unidirectional transport of ATM cells. The cells carried over a specific VC are identified by a Virtual Channel Identifier (VCI).

- Virtual Path (VP)—a term used to describe a bundle of VCs. The VCs associated with a VP are transported over a transmission path within the network as a group. Each bundle of VCs is identified by a Virtual Path Identifier (VPI).

The VPI and VCI values are carried in the header of the ATM cell and, as we will see later in this chapter, these values are used for routing the cell across the ATM network. The relationship between virtual path, virtual channel, and transmission path is shown in Figure 2-1.

The Physical Layer is subdivided into three levels that represent the transmission levels supported within the network:

- Transmission Path—represents a logical connection between the point at which information is assembled into a standard frame format for transport over a transmission medium at a given data rate, and the point at which the frame format is disassembled. Examples of frame formats for ATM are SONET, SDH, and PDH [2][3][4]. The transmission path is commonly referred to by the term *path*. Network Elements that exist at the ends of a transmission path are referred to as *Path Terminating Equipment* (PTE).

- Digital Section—a transmission medium, together with associated equipment, required to provide a means of transporting information between network elements, one of which originates

Figure 2-1.
VP, VC, transmission path relationship.

the line signal, and the other, which terminates the line signal. Network Elements that may terminate digital sections have the capability of multiplexing or demultiplexing traffic into a composite signal with a higher or lower data rate. Other terms for Digital Section include *Multiplex Section* (SDH) and *Line* (SONET). The Network Elements are referred to as *Line Terminating Equipment* (LTE). Note that a digital section may exist between two LTEs, two PTEs, or between a LTE and a PTE.

- Regenerator Section—a transmission medium, together with associated equipment, required to regenerate a signal on the medium. The only capability required of network elements that terminate regenerator sections is the ability to regenerate a signal. The network element need not have the ability to modify the content of the signal. In SONET systems, this is merely called a *section*. Network elements that provide this capability are called *Section Terminating Equipment* (STE). A section may exist between any pair of networks elements (PTE, LTE, STE).

We can now construct a hierarchical model showing the conceptual relationship between the ATM Layer and Physical Layer. This is shown in Figure 2-2, which introduces two new concepts: Virtual Path Connection (VPC) and Virtual Channel Connection (VCC). These will be discussed further in the next section.

2.1 The ATM Switching Concept

As mentioned earlier, the ATM Layer consists of VC and VP sublayers. These sublayers form the basis of switching within ATM networks. Correspondingly, there are the two levels of ATM connection:

- Virtual Path Connection (VPC)—extends from the point at which virtual channels are assigned VCI values and associated with a virtual path, to the point at which virtual channels are removed from the virtual path or have their VCI values modified. These points are referred to as the VP connection endpoints.

- Virtual Channel Connection (VCC) — extends between points where adaptation layer functionality is performed. The adaptation

Chapter 2: ATM Technology Concepts

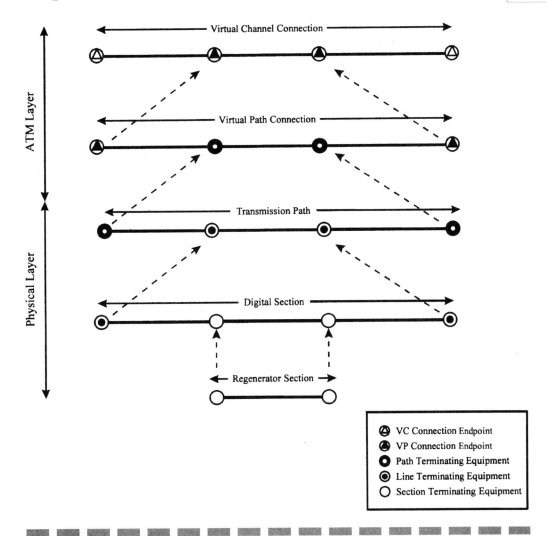

Figure 2-2. ATM transport architecture hierarchical structure.

layer for ATM will be discussed in further detail in the next chapter, but the basic function of the adaptation layer is to format data for transport over an ATM connection on the sending end of the connection. On the receiving end the adaptation layer extracts data from the ATM connection for presentation to a higher layer protocol which is to use the data. These points are the VC connection endpoints.

VP and VC connections are established by either of the following methods:

- Permanent Virtual Connection (PVC)—These are connections that are established, often by a network operator using manual procedures. Once established, the connection is maintained until manually torn down.
- Switched Virtual Connection (SVC)—These connections are established on demand by signalling procedures between connection endpoints. Under normal circumstances, these connections are maintained until terminated by the communicating parties.

2.1.1 Virtual Path Switching

Virtual path switching is shown in Figure 2-3. In this figure, virtual paths (containing virtual channels) identified by a VPI enter an ATM Network Element on one side and exit on the other with a different VPI. For example, the virtual path entering the network element with VPI=1, exits with VPI=6. This is called *virtual path switching*.

Note, however, that in each case of VP switching, the bundle of VCs within each VP remain the same even though the VPI has been changed (or *translated*). Since none of the VCs within the virtual path have been removed or have had their VCI values translated, the VPC remains the same even though the VPI value has been translated. This introduces additional ATM connection concepts:

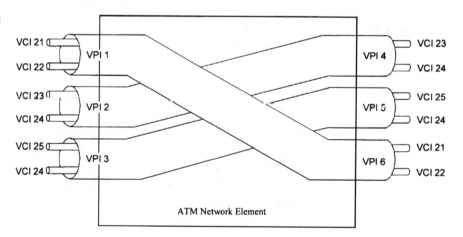

Figure 2-3.
Virtual path switching.

- Virtual Path Connection Identifier (VPCI)—The VPCI is a value which remains constant throughout the VPC. Thus, in our example above, the virtual path with VPI=1 being translated to VPI=6 at a network element, would have the same VPCI value. The VPCI is not used for routing of individual ATM cells within a network and is therefore not carried in the ATM cell header. The VPCI is used, for instance, during VPC establishment, or to identify the VPC for network management.
- Virtual Path Link (VPL)—identifies a section of a VPC that has a single VPI value. Thus, the virtual path in our example above consists of two virtual path links: one link with VPI=1, and a second with VPI=6.
- Connecting Point—the point at which links at the same layer in the ATM transport hierarchy are connected. In Figure 2-3, the network element is the connecting point between virtual path links.

2.1.2 Virtual Channel Switching

Virtual channel switching is shown in Figure 2-4. In this figure there is translation of both VPIs and VCIs. The latter case is called *virtual channel switching*. Virtual channels associated with virtual path VPI=1 have their VCI values translated within the network element. The point at which a virtual channel has its VCI value translated corresponds to the termination of the associated virtual path. This is called a *Virtual Path Connection Endpoint*. Figure 2-4 also shows virtual path VPI=4 having its VPI value translated to VPI=5 within the same network element. However, since within the VP no VCs are added, removed, or have VCI values changed, there is no VPC endpoint within the network element for this virtual path.

There is no *Virtual Channel Connection Endpoint* shown in Figure 2-4. What the figure does show are Virtual Connection Links. The VCCs entering the network element over VPI=1 consist of two virtual connection links: for example, one of the VCCs consists of virtual channel links (VPI=1, VCI=21) and (VPI=2, VCI=24).

3. B-ISDN Network Architecture

Next we will discuss the elements in a B-ISDN network which carries ATM traffic. As shown in Figure 2-5, a B-ISDN consists of two components [5]:

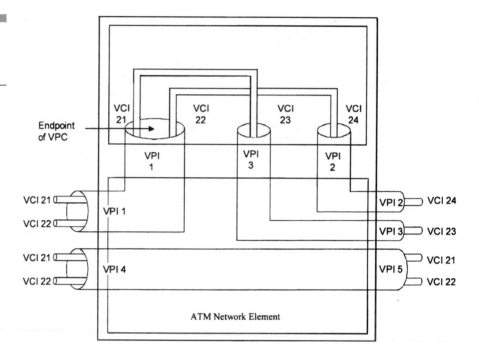

Figure 2-4. Virtual path and virtual channel switching.

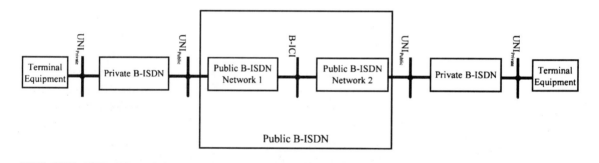

Figure 2-5. B-ISDN reference configuration for mixed private public B-ISDN.

1. Customer Premises Equipment (CPE), which includes Terminal Equipment (TE) and other customer network equipment. Customer network equipment may also include one or more ATM network elements configured to form a Private B-ISDN.

2. The Public B-ISDN consists of one or more public carrier B-ISDNs.

Figure 2-5 also shows the locations of significant reference points in the network [6].

- Private User-Network Interface (UNI)—defines an interconnection point between TE and switching ATM network elements which are operated and managed within a customer premises network.
- Public UNI—defines an interconnection point between CPE and a network element in the public B-ISDN carrier network.
- Network Node Interface (NNI)—defines an interconnection point between ATM switching elements within the (Public or Private) B-ISDN (not shown in Figure 2-5).
- Broadband Inter-Carrier Interface (B-ICI)—a special case NNI which defines the interconnection point between network elements within different public carrier B-ISDN.

TE that attach directly to a UNI are assumed to be able to interface directly to a B-ISDN as a VC connection endpoint. For TE that cannot perform this function an intermediate Terminal Adaptation (TA) function is required which adapts a non-B-ISDN TE to the B-ISDN and performs VC connection termination.

3.1 Reference Configuration at the UNI

While documents from the ATM Forum use the terms "Public UNI" and "Private UNI" to describe customer premises ATM interfaces, ITU Recommendations use terminology inherited from ISDN. Although the argot between the two organizations are not identical in the semantic sense, they are substantially analogous in concept. The ITU UNI model, shown in Figure 2-6, consists of functional groups and reference points. The functional groups are the following:

1. B-NT1—Provides line interface and termination to the public carrier B-ISDN. Equivalent to ISDN NT1.
2. B-NT2—Equivalent to ISDN NT2, performs functions such as cell delineation, multiplexing/demultiplexing, and switching. Many customer premises ATM switching elements provide combined B-NT1 and B-NT2 functionality. The Private B-ISDN of Figure 2-5 is an example of this combined functionality.
3. B-TE1—TE that is able to interface directly to a B-ISDN.
4. B-TE2—TE that is not able to interface directly to a B-ISDN.

Figure 2-6.
UNI reference configuration.

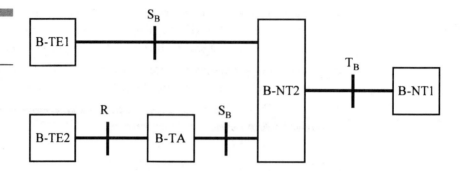

5. B-TA—Provides functionality that allows a B-TE2 to access a B-ISDN. Some ATM network elements provide B-TA functionality.

The reference points are the following:

- T_B—This is the reference point at which a B-NT2 interfaces to a B-NT1. There is one T_B reference point per B-NT1. This reference point is equivalent is the Public UNI shown in Figure 2-5.
- S_B—This is the reference point at which a B-TE or B-TA interfaces to a B-NT2. There may be one or more S_B reference points per B-NT2. This reference point is equivalent is the Private UNI shown in Figure 2-5. For a TE that is directly connected, the Public B-ISDN, the T_B, and S_B reference points are coincident.
- R—This is the reference point at which a B-NT2 interfaces to a B-TA.

3.2 ATM Network Elements

Network elements in ATM equipment can be described by the following categories [7][8]:

1. Cross-Connects: these devices provide connectivity between any input port and any output port. Cross-connects provide PVC connectivity only. Cross-connects are also subdivided into VP cross-connects, which support connection at the VP level only (such as in Figure 2-3), and VC cross-connects, which provide connectivity at either the VP or VC level (as shown in Figure 2-4).
2. Switches: add support for SVC connectivity to the capabilities supported by cross-connects. Switches are also categorized as VP switches and VC switches.

3. Interworking Cross-Connects: these devices provide an integrated TA function to allow non-B-ISDN-capable TE to interface to B-ISDN. Otherwise the Interworking Cross-Connect provides the same functionality as a Cross-Connect described in (1) above.
4. Interworking Switches: provide the functionality of switches described in (2) above with an integrated TA function.

3.3 ATM Switching—An Example

The basic mechanism for ATM switching within a VC switching element is characterized by the following steps:

1. A cell arrives on an input port at the ATM switching element with values (VPI_{in}, VCI_{in}) in the cell header.
2. The input port and ATM header information are used to locate and entry in a translation table. This table contains information used in ATM cell routing.
3. The located entry contains values (VPI_{out}, VCI_{out}, output port#).
4. The (VPI_{in}, VCI_{in}) values in the ATM header of the received cell are "translated" to (VPI_{out}, VCI_{out}).
5. The cell is sent out of the switching element on the output port as specified in the translation table with the translated cell header.
6. Subsequent ATM switching elements repeat the process until the cell arrives at the connection endpoint.

An example illustrating this mechanism is presented in Figures 2-7, 2-8, and 2-9. The example discussed here does not necessarily represent the way that this function is actually implemented in any ATM switch. In this example, the lookup table can be viewed as a three-dimensional matrix of entries with VPI, VCI, and Input Port making up the axes of the matrix. Each entry in the matrix contains a (Output Port, Output VPI, Output VCI) tuple, in addition to other connection-related information. In Figure 2-8, (VPI_{in}, VCI_{in}) are translated by the following procedure:

1. The input port number I1 is used to select a two-dimensional "slice" of the overall matrix. This "slice" consists of the possible values for VPI and VCI for that input port (Figure 2-8a).

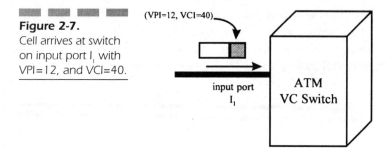

Figure 2-7.
Cell arrives at switch on input port I_1 with VPI=12, and VCI=40.

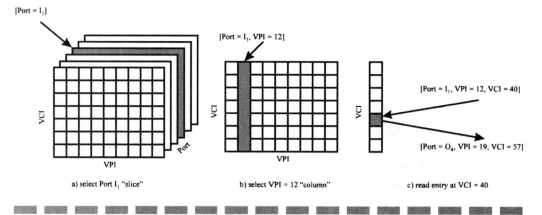

a) select Port I_1 "slice"

b) select VPI = 12 "column"

c) read entry at VCI = 40

Figure 2-8. Table lookup process.

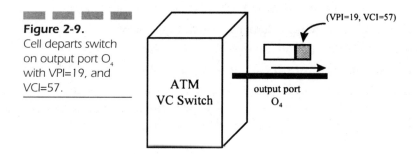

Figure 2-9.
Cell departs switch on output port O_4 with VPI=19, and VCI=57.

2. Given the "slice" from (1), the VPI_{in} value is used to select a column (Figure 2-8b).

3. Next, the VCI_{in} value is used to select an entry (Figure 2-8c). This entry contains the following: (Output Port = O4, VPI_{out} = 19, VCI_{out} = 57).

4. Finally, the cell departs the ATM switch on output port O4. The ATM cell header contains VPI = 19, and VCI = 57 (Figure 2-9).

This example also helps to illustrate the difference between VP and VC switching elements in terms of complexity. At the UNI, an ATM cell is able to select 256 unique virtual paths and 65,536 unique virtual circuits within each of those VPs. Thus, for a VC switching element to be able to accommodate the entire range of VPI/VCI values would require enough memory in the element to store a table of 16,777,216 elements for each port. Even a relatively modest 8-port switching element would have to be able to store a staggering 134,217,728 entries. By comparison, a VP switching element need only maintain storage for 256 entries per port, or a more modest 2,048 entries for an 8-port switching element.

In addition to the hardware costs of such a VC switching element (switch or cross-connect), the administrative costs associated with establishing and maintaining connections on a cross-connect of this type would be prohibitive. In reality, however, it is not practical for ATM switching elements to be able to store information for every possible ATM connection. Instead, most VC switching elements support only a subset of the possible VPI and VCI ranges at each port.

3.4 UNI Connection Topologies

3.4.1 Point-to-Point Topologies

Consider the following scenarios, in which we will assume that three separate subscriber locations are being connected across a Public B-ISDN:

- VC Level Connection: In this case (Figure 2-10), individual VCCs are passed across the UNI and are switched within the Public B-ISDN. In a PVC-only network, this requires that the subscribing customers contact the Public B-ISDN service provider to manually establish a new connection each time additional connections are to be created between subscriber locations. In an SVC environment, connections are established and created on-demand without manual intervention by the service provider.

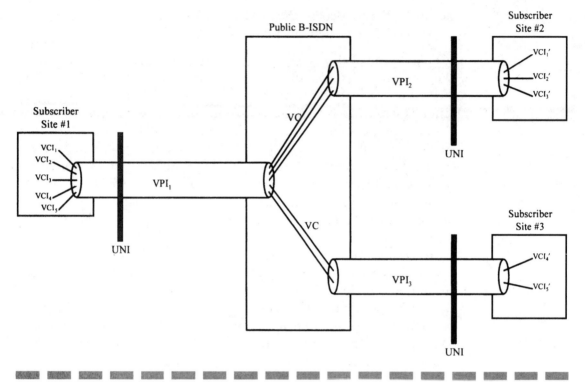

Figure 2-10. VC switching with public B-ISDN.

- VP Level Connection: In this scenario (Figure 2-11), individual VPCs are passed across the UNI with the Public B-ISDN switching at the VP level only. Here, VC switching is performed within the Customer Private Network only. This allows the subscriber to establish new VCCs between sites connected by one or more virtual paths without service provider intervention.

3.4.2 Point-to-Multipoint Topologies

In addition to the point-to-point topologies between individual TEs discussed above, B-ISDN is also defined to support multipoint connection topologies [9]. These topologies are based upon a Multipoint Connection Point (MPCP), which serves as a reference point that binds a port to a set of connections. The following types of multipoint connections may then be defined:

Chapter 2: ATM Technology Concepts

- Broadcast—The MPCP binds an output port on an ATM network element ("root") to a set of unidirectional connections to one or more other ATM network elements ("leafs") (Figure 2-12a). The MPCP transmits identical cell traffic streams over each connection.
- Merge—The MPCP binds an input port to a set of unidirectional connections (Figure 2-12b). The MPCP merges cell traffic streams received over each of the connections.
- Bidirectional Composite Multipoint—The MPCP binds a bidirectional port to a set of bidirectional connections (Figure 2-12c). This gives the MPCP the composite capabilities of Broadcast and Merge.
- Bidirectional Full Multipoint—Consists of a set of network elements, each with Bidirectional Composite Multipoint capability (Figure 2-12d).

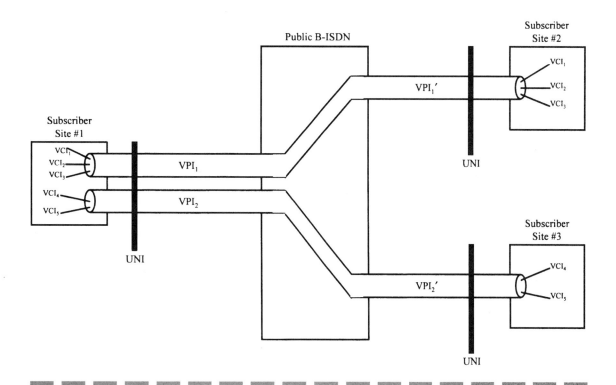

Figure 2-11. VP-only switching within public B-ISDN.

Figure 2-12.
Multipoint connection configurations.

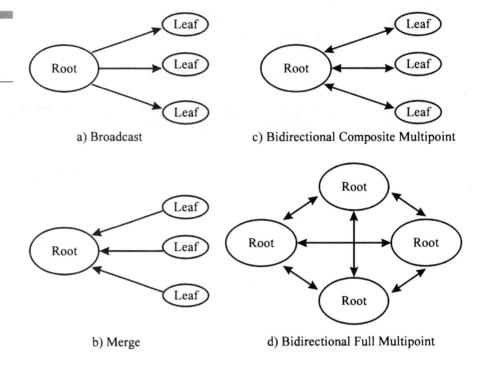

a) Broadcast

b) Merge

c) Bidirectional Composite Multipoint

d) Bidirectional Full Multipoint

4. Quality of Service

The translation table referred to in the previous section also contains information pertaining to another important feature of ATM—the Quality of Service (QoS) to be provided to cell traffic carried over the connection.

A "service" with communication networks may be described at two levels:

- Teleservices are services that are visible to end users at the T interface. Examples of teleservices include voice telephony, videoconferencing, fax, and data [10].
- Bearer Services are the underlying attributes within the B-ISDN that provide the capability for the transmission of teleservices between T interfaces [11].

In ATM networks the QoS is defined by bearer service attributes [12][13]. The basis for QoS is a set of performance objectives that are to be met by the B-ISDN. These performance objectives are to be met whether the end-to-end connection traverses one or more public carrier networks. QoS parameters that may vary on a per-connection basis include [14]:

1. Cell Loss Ratio (CLR): This is the probability that a cell is successfully transmitted end-to-end across the network.

2. Cell Transfer Delay (CTD): This is the transit time required for a cell to traverse the network.

3. Cell Delay Variation (CDV): This measures the variation in CTD among cells carried over a connection.

Different teleservices have different QoS requirements resulting in different classes of QoS. For example, telephony and other isochronous teleservices may specify a QoS class that may require low CTD and (ideally) zero CDV. A QoS class for data servers, however, may allow higher CTD and CDV.

The QoS class for a connection must be determined at connection establishment. In addition to selection of the QoS class there must be a specification of a set of parameters which describe the traffic characteristics of the source. This set of parameters is referred to as a *traffic descriptor* [15][16]:

- Peak Cell Rate (PCR): the maximum rate at which the source will transmit cells over the network.
- Sustained Cell Rate (SCR): the rate at which the source will typically transmit cells over the network. When specified, this parameter is less than the PCR.
- Maximum Burst Size (MBS): specified when an SCR is declared, this parameter defines the maximum number of cells that a source may transmit cells over the network at the PCR.

The subscriber chooses a QoS appropriate for the cell traffic pattern that is required for the associated teleservice. Traffic descriptors can describe a widely varying range of cell traffic patterns, but in principle, traffic characteristics fall into two broad categories:

- Constant Bit Rate (CBR)—Cell traffic is transmitted from the source at a constant rate with a fixed interval of time between cell emissions. The traffic descriptor for this type of pattern typically consists of a PCR specification only.
- Variable Bit Rate (VBR)—Cell traffic is transmitted from the source with a variable interval between cell emissions. This includes periods where there are no cell emissions, and other

periods where bursts of cells are emitted at a high rate. The traffic descriptor for VBR consists of specifications for PCR, SCR, and MBS. The VBR subscriber is able to transmit cells at rates up to the SCR at any time, but can only transmit bursts of up to MBS number of cells at the PCR.

In addition, the choice of QoS by a subscriber is influenced by factors such as the pricing structure and set of services offered by the available public carriers.

Ultimately, the QoS for a connection translates into how ATM switching elements within the network prioritize cells from that connection relative to cells from other connections. For instance, during the table lookup process shown in Figure 2-8, the located entry could contain information that allows the switch to determine how soon to send the cell on the outgoing link relative to other cells which are also to be sent over the same link. If the cell has the highest relative priority, it will be sent on the outgoing link first, regardless of its time of arrival at the switch relative to the other cells; if it has the lowest priority, then it will be sent last.

The specific mechanisms for providing differential treatment of cells from connections of varying QoS are an implementation matter. As such, this becomes a basis for competitive differentiation among ATM switch vendors, as different ATM switching elements may implement different methods of prioritizing cell traffic by QoS.

For public B-ISDN, the agreed-upon QoS constitutes a commitment from the public carrier for traffic transmitted across the T_B interface into the network for delivery to a subscriber across the T_B interface on the other end of the connection. While public carriers are not responsible for private B-ISDN, the end-to-end QoS must include commitments from any private networks that exist along the connection path. The subscriber's responsibility is to transmit a traffic pattern that is in conformance with the traffic descriptor while network operator's responsibility is to ensure that all subscribers receive the committed QoS independent of the actions of other subscribers. This requires that the network operators have the means to enforce conformance among subscribers, since failure to do so would allow some subscribers, either unintentionally or maliciously, to unfairly utilize more than their share of network resources. This, in turn, could degrade the QoS received by other subscribers.

This enforcement mechanism is called Usage Parameter Control (UPC). UPC consists of a set of actions to verify that transmitted cells are in compliance with the appropriate traffic descriptor at the Public UNI. Since UPC is performed on a per connection basis the information necessary to monitor the conformance of a connection is also retrieved during the table lookup process depicted in Figure 2-8. Cell traffic that is in compliance is treated according to the committed QoS. The network operator, however, is not committed to provide this QoS for cell traffic that is not in compliance with the traffic descriptor. In this case, the network may take corrective actions that include the following:

i. Discard the cell. This eliminates the cell from the network and raises the CLR for the connection.

ii. Modify the ATM cell header to lower the delivery priority for the cell. This effectively raises the CLR by making it less likely that the cell will be delivered to the receiving party.

Since individual public carriers manage traffic on their networks independently, there is also the need for a similar function between public carriers. This is called Network Parameter Control (NPC). NPC is enforced at the B-ICI.

Quality of Service is discussed in further detail in Chapter 5, Traffic Management.

5. ATM Signalling: On-Demand Connection Establishment

Signalling provides the capability for establishment of SVCs between communicating endpoints on demand by the parties involved in the connection [1]. The entities involved in requesting connection establishment and disconnection are referred to as *signalling endpoints*. The mechanism for SVC establishment is the exchange of signalling messages over a dedicated VCC, which identifies the calling and called parties and includes the QoS and other bearer service capabilities required for the connection [6][17]. This exchange of messages constitutes a negotiation between the connection endpoints, which is initiated when a calling party sends a signalling message requesting establishment of an SVC to a called party. The called party may do either of the following:

- Accept the message, thereby establishing a connection.
- Reject the message, resulting in a failed call attempt.

- Send a return message to offer an alternative set of connections parameters. The calling party may accept or reject this message.

The network may reject signalling messages from either connection endpoint without negotiation. For example, the network may not be able to commit to the requested QoS.

There are two scenarios for signalling between subscribers across a public network:

1. User-to-Network — In this case a signalling endpoint within Customer_A premises uses a signalling VCC to communicate across the UNI with a corresponding signalling endpoint within the public ATM network to request establishment of an SVC to Customer_B (Figure 2-13). This network signalling endpoint (or a peer signalling endpoint within the public ATM network) communicates with the Customer_B signalling endpoint on behalf of the Customer_A signalling endpoint. After successful negotiation between the connection endpoints, an SVC is established which connects Customer_A and Customer_B. The network assumes responsibility for selecting a route through the public network.

2. User-to-User-Here a signalling endpoint within Customer_A premises uses a signalling VCC to communicate directly with a signalling endpoint within the Customer_B premises. This exchange is transparent to the public ATM network. In this scenario, the public ATM network performs VP-only switching (such as shown in Figure 2-11) between the customer premises, while user-to-user signalling performs SVC establishment of VCCs within the VPC (Figure 2-14).

The signalling configuration described above is referred to as a point-to-point signalling configuration in which the UNI serves a single signalling endpont. In this case, signalling between connection endpoints across the UNI uses a well-known VCI value. A second signalling configuration is called *meta-signalling* [18]. In meta-signalling there are multiple signalling endpoints served by a common UNI on the subscriber side. In the meta-signalling protocol the network assigns a signalling VCC to each signalling endpoint individually. At the time of this writing, meta-signalling is not supported by ITU recommendations or ATM Forum standards for signalling.

Chapter 2: ATM Technology Concepts

Figure 2-13.
User-to-network signalling.

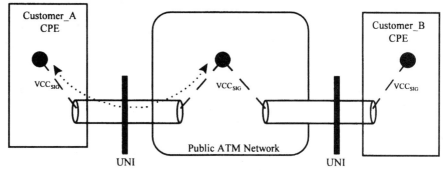

a) Customer_A requests SVC VCC establishment to Customer_B through Public ATM Network over Signalling VCC

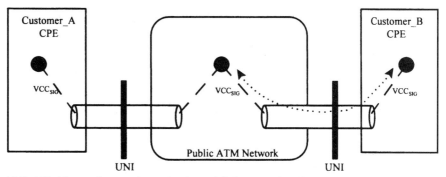

b) Public Network negotiates SVC establishment with Customer_B over Signalling VCC

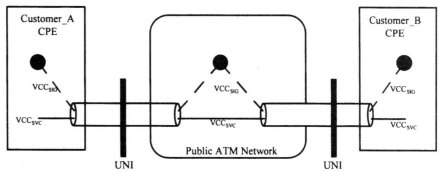

c) SVC VCC established between Customer_A and Customer_B

● Signalling Connection Endpoint

Figure 2-14.
User-to-user signalling.

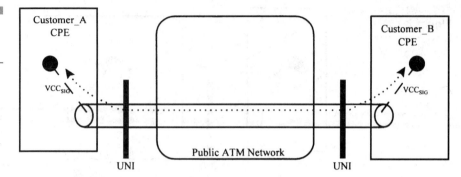

a) Customer_A negotiates SVC establishment directly with Customer_B over Signalling VCC

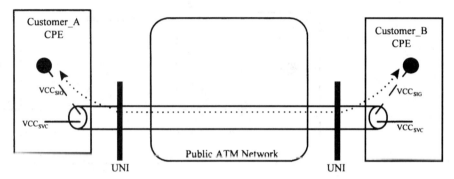

b) SVC VCC established between Customer_A and Customer_B

6. Managing the Network

Operations Administration, Maintenance, and Provisioning (OAM&P) consists of the set of tasks which provide for overall management of the ATM communications network. These consist of OAM tasks [19]:

- Performance Monitoring— a function that measures the performance of the network in providing end-to-end cell delivery;
- Fault Detection—functions that detect and report failures in ATM network elements, connections, or transmission facilities;

and provisioning tasks:

- Configuration Management—which supports configuration of network elements and PVC establishment.

Chapter 2: ATM Technology Concepts

6.1 ATM Network Management Reference Model

The conceptual model for management of ATM networks is based upon the Telecommunications Management Network (TMN) as defined by the ITU and shown in Figure 2-15 [20]. The TMN model defines network management on five levels:

1. Business Management Layer (BML). The business management layer has responsibility for the total enterprise and is the layer at which agreements between carriers are made.

2. Service Management Layer (SML). Service management is concerned with and responsible for the contractual aspects of

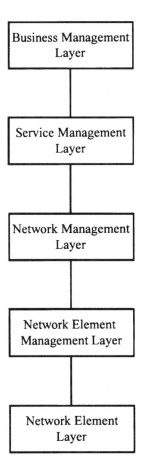

Figure 2-15.
TMN functional hierarchy.

services that are being provided to customers or are available to potential new customers.

3. **Network Management Layer (NML).** This layer has responsibility for all network elements in the network, both individually and as a set. The network management layer interacts with these network elements through the element management layer and is not concerned with how a particular network element provides services internally.

4. **Element Management Layer (EML).** The element management layer manages each network element on an individual basis and provides an abstraction of the services provided by that element to higher layer entities.

5. **Element Layer (EL).** This layer refers to the network element itself and contains functions related specifically to the technology, vendor, and resources that provide basic communications services.

The ATM network management model as defined by the ATM Forum is shown in Figure 2-16 [21][22]. This is a distributed model that consists of a Network Management Systems (NMS) for the Private (Customer Premises) Network and an NMS for each of the independently managed Public Carrier Networks. The NMS, and the underlying network elements managed by the NMS, provide the functionality defined by the lower 3 layers of the TMN model.

The ATM network management model also defines a set of interfaces that allow the respective NMS to manage its respective network elements and to communicate with peer NMS:

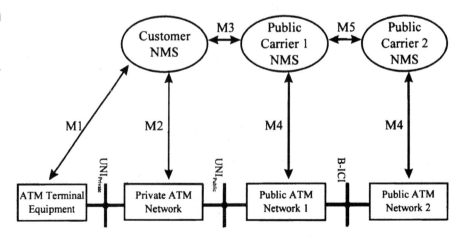

Figure 2-16.
ATM network management model.

- M1 is the ATM management interface needed to manage ATM terminal equipment.
- M2 is the management interface needed to manage a private ATM network.
- M3 is the network management interface that allows a customer to manage the resources that they use in a public ATM network. The public carrier may limit capabilities over this interface to only allow the customer to monitor the public network and to retrieve status information only. There may be more than one M3 interface for a customer using the services of multiple public carriers.
- M4 is the ATM management interface that allows a public carrier to manage its own public ATM network. This NMS may also include the BML and SML functions.
- M5 is the management interface that supports interaction between the NMS of different ATM public network providers.

In relation to the TMN model, the M1, M2, and M4 interfaces are equivalent to the TMN Q_3 interface, while the M3 and M5 interfaces are equivalent to the TMN X interface.

It is important that network management interfaces be separate from the transmission facilities that they monitor and control. In this way, the network may continue to be managed even if the underlying transmission facilities should fail or otherwise be removed from service.

6.2 Management Information Base

The basic network management entity is the *managed object*. A managed object is an abstraction of a network resource that represents its properties to an NMS. Of particular significance is the relationship between these properties and the operational behavior of the network resource. A network resource may be viewed as any physical or logical entity whose behavior influences communications over the network. In Figure 2-2, examples of physical and logical entities include the following:

- Physical entities: section, line, or path terminating equipment such as ATM switches, cross-connects, multiplexers, line interfaces, and transmission media;
- Logical entities: virtual path connections, virtual channel connections, VP and VC connecting endpoints, transmission, digital, and regenerator sections.

A collection of managed objects forms a *management information base* (MIB). A set of managed objects associated with a particular network element would be directly managed by an element management system. On the other hand, managed objects that span multiple network elements would be directly managed by a network management system. For example, management of a port on an ATM switch is an example of an EML function, while management of an end-to-end ATM connection is an example of an NML function.

6.3 Network Management Activities

In this section we will give examples of the principles of ATM network management activities. Further detail on OAM&P may be found in Chapter 6, Operations Administration, Maintenance, and Provisioning.

6.3.1 Configuring Network Elements

There are numerous configuration tasks required to install and configure an ATM network element from a network management station [23]. For the most part, these tasks are not generalizable but are vendor-specific. However, configuration of a UNI or NNI would include the following tasks:

1. Assign a unique interface identifier.
2. Define the maximum number of VPCs that are to be supported on the interface.
3. Define the maximum number of VCCs that are to be supported.
4. Establish the maximum aggregate bandwidth for all connections carried over the interface.

6.3.2 Permanent Virtual Circuit Establishment

A network management station would configure the PVC based upon the following:

Chapter 2: ATM Technology Concepts

1. Identification of VP or VC connection endpoints. This identification could be based upon the interface identifiers associated with the connection endpoints.
2. Determination of whether the connection is unidirectional or bidirectional.
3. Definition of traffic descriptors for the connection. For a bidirectional connection, separate traffic descriptors are established for each direction.

The NMS may be responsible for determining a route through the network that connects the connection endpoints and for communicating with the individual EML systems that configure the corresponding ATM network elements to provide the requested end-to-end service. In a customer-premises NMS, the subscriber has control over routing within the Private ATM network over the M2 interface, but not within the public networks. If configuration control is available, the customer may request PVC connection establishment over the M3 interface. Alternatively, for the customer with end-to-end VPC service through the public network (as shown in Figure 2-11), a customer NMS allows the customer to establish individual end-to-end VCCs. Similarly, individual public carriers do not allow other carriers to control routing within their networks. If an M5 interface is available between public carriers, a carrier may request PVC establishment through another carrier's network.

The NMS is also able to modify parameters for existing PVCs or to terminate existing connections.

6.3.3 Collection of Traffic Performance Statistics

One of the tasks of performance monitoring is to collect statistics on the number of cells sent from one point in the network and the number of those cells that are received at another point. Statistics may be collected over an end-to-end connection or at points in between.

The basic mechanism relies on each node in the network maintaining statistics on the number of cells it transmits over a connection, and the number of cells that it receives. The translation table entry from the example illustrated in Figure 2-8 would also contain statistics on the number of cells received on the input port from the connection. If this is a bidirectional connection, there will be a second translation table entry (with a different port_id/VPI/VCI combination) that will contain statistics on the number of cells transmitted from the node in the opposite direction.

Within an ATM network, a network element, Node_A, will have statistics on how many cells it transmits over a connection, but it does not independently have statistics on how many cells were received by a downstream network element, Node_B. To make this determination, the NMS must configure Node_A and Node_B (which may represent any two ATM network elements along the end-to-end connection path) to correlate traffic information, which may then be retrieved by the NMS.

The statistics collection mechanism is shown in Figure 2-17, in which Node_A sends cells to Node_B over an ATM connection at regular intervals. These cells are different from the cell traffic that supports customer teleservices because the specific purpose of this cell, referred to as an OAM cell, is to collect statistics on network performance. A cell sent from Node_A contains information on how many cells have been transmitted over the connection since the last OAM cell was sent (Figure 2-17a). When received by Node_B, it sends a return cell to Node_A containing the number of cells that were received over the ATM connection at node B over that interval (Figure 2-17b). Upon receipt of this return information, a CLR statistic for the ATM connection at Node_A may be computed.

Figure 2-17. Performance monitoring.

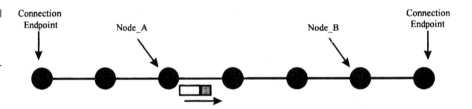

a) Performance Monitoring OAM Cell from Node_A to Node_B

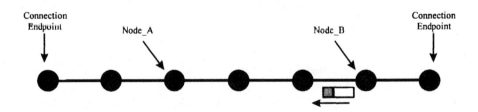

b) Return OAM Cell from Node_B to Node_A

6.3.4 OAM Cell Loopback

When a customer requests that a PVC be established, the ability of the ATM network to deliver cell traffic depends in part upon the accuracy with which the connection was established among each of the network elements along the end-to-end connection path. This task may be complicated when the connection spans more than one public carrier network. One method of verifying the accuracy of PVC establishment is through OAM cell loopback. OAM cell loopback provides an NMS with the ability to send a cell from Node_A to Node_B, which upon arrival at Node_B is looped back and returned to Node_A. When the two points involved in the loopback are connection endpoints, this capability is useful in verifying that the end customers are able to communicate over the connection. If the connection was not properly established, loopback testing between intermediate nodes may be used to locate where the connection path is broken.

6.3.5 Fault Detection, Reporting, and Recovery

There are times when failures occur within the network. The source of the failure may be due to hardware, software, or both. In any case, there must be a mechanism for detecting when a failure occurs, where it occurred, and the cause of the failure. In turn, this information must be relayed to an NMS so that proper corrective action may be taken to recover from the failure to allow the network to resume service to customers at the committed QoS.

Figure 2-18 illustrates the impact on a bidirectional VCC due to failure of an output port on ATMswitch_A. When ATMswitch_B detects that it is no longer receiving cell traffic on the VCC, it sends an OAM cell toward Customer_B, which indicates that a failure has occurred on the connection and that it can no longer receive cell traffic from Customer_A. This OAM cell also indicates that the failure was detected at ATMswitch_B (we will assume that the connection from Customer_B to Customer_A is unaffected). For a VCC, this cell is called a VC-AIS (Alarm Indication Signal) cell.

ATMswitch_A detects that a failure has occurred on one of its output ports and sends an OAM cell on the reverse path VCC toward Customer_A which indicates that it can no longer send cell traffic to Customer_B. This OAM cell indicates that the failure was detected at ATMswitch_A.

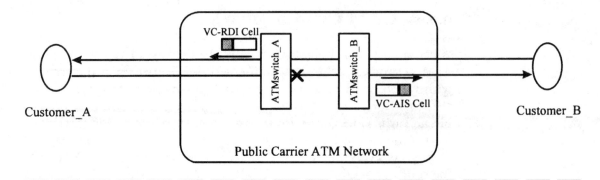

Figure 2-18. Fault detection on a VCC: Failure detected at ATMswitch_A.

This cell, referred to as a VC-RDI (Remote Defect Indication) cell, indicates that the failure has not occurred on the VCC on which the VC-RDI is being received, but on the VCC in the opposite direction.

These OAM cells provide information that is reported to the NMS to assist in locating the source of the trouble at the network level. The Customer_A and Customer_B NMS, as well as the NMS of any other public carriers involved in the VCC between Customer_A and Customer_B, will receive AIS or RDI notification. Depending upon the failure detection capabilities of the affected network element, the EML system that manages ATMswitch_A would also receive a failure report with specific information about the failure of the affected output port.

The next task is to recover from the failure and to take action to restore full service on the network. In some cases, manual intervention is required in which service technicians are dispatched to the suspected source of trouble to effect a repair. If this is the sole means of fault recovery, the service outage could last for a considerable length of time. If possible, a network manager at an NMS may configure an alternate path around the failure to restore service. Software fault recovery seeks to automate this process and to affect recovery actions with minimal service impact.

References

1. B-ISDN General Network Aspects, International Telecommunication Union, Recommendation I.311, March 1993.

2. Synchronous Optical Network (SONET)—Basic Description including Multiplex Structure, Rates, and Formats, American National Standards Institute, ANSI T1.105, 1995.

3. Network Node Interface for the Synchronous Digital Hierarchy (SDH), International Telecommunication Union, Recommendation G.707, March 1996.

4. ATM Cell Mapping Into Plesiochronous Digital Hierarchy (PDH), International Telecommunication Union, Recommendation G.804, November 1993.

5. B-ISDN User-Network Interface, International Telecommunication Union, Recommendation I.413, March 1993.

6. ATM User-Network Interface Specification, Version 3.1, ATM Forum, September 1994.

7. Types and General Characteristics of ATM Equipment, International Telecommunication Union, Recommendation I.731, March 1996.

8. Handel, Rainer, Manfred Huber, and Stefan Schroder, *ATM Networks*, Addison-Wesley, 1994.

9. Functional Architecture of Transport Networks Based on ATM, International Telecommunication Union, Recommendation I.326, November 1995.

10. Definition of Teleservices, International Telecommunication Union, Recommendation I.240, 1988.

11. Definition of Bearer Service Categories, International Telecommunication Union, Recommendation I.230, 1988.

12. Broadband Connection-Oriented Bearer Service, International Telecommunication Union, Recommendation F.811, July 1996.

13. Broadband Connectionless Data Bearer Service, International Telecommunication Union, Recommendation F.812, August 1992.

14. B-ISDN ATM Layer Cell Transfer Performance, International Telecommunication Union, Recommendation I.356, October 1996.

15. Traffic Control and Congestion Control in B-ISDN, International Telecommunication Union, Recommendation I.371, August 1996.

16. Traffic Management Specification, Version 4.0, ATM Forum, April 1996.

17. Broadband Integrated Services Digital Network (B-ISDN)—Digital Subscriber Signalling System No. 2 (DSS 2—User-Network Interface (UNI) Layer 3 Specification for Basic Call/Connection Control, International Telecommunication Union, Recommendation Q.2931, February 1995.

18. B-ISDN Meta-Signalling Protocol, International Telecommunication Union, Recommendation Q2120, February 1995.
19. B-ISDN Operation and Maintenance Principles and Functions, International Telecommunication Union, Recommendation I.610, November 1995.
20. Principles for a Telecommunications Management Network, International Telecommunication Union, Recommendation M.3010, October 1992.
21. Customer Network Management (CNM) for ATM Public Network Service (M3 Specification), Revision 1.05, ATM Forum, January 1996.
22. M4 Network-View Interface Requirements and Logical MIB, Document AF-NM-0058.000, ATM Forum.
23. Asynchronous Transfer Mode Management of the Network Element View, International Telecommunication Union, Recommendation I.751, March 1996.

PART 2

CHAPTER 3

ATM Communications Protocols

1. Introduction

ATM provides a mechanism for communication between two end points within the context of a generalized protocol between the communicating entities. One example of such a communications protocol is the Open Systems Interconnection (OSI) reference model [1]. This model consists of seven layers, each layer providing a set of services to the next higher layer:

1. Physical Layer—Concerned with transmission of a bit stream over a physical communications medium, including mechanical, electrical, functional, and procedural characteristics for access to the physical link.
2. Data Link Layer—Provides reliable transfer of information across the physical link consisting of blocks that include data, synchronization, error control, and flow control information.
3. Network Layer—Responsible for establishing, maintaining, and terminating connections while providing upper layers with independence from the data transmission and switching technologies used to connect end systems. The end-to-end connection between communicating applications supported by this layer may span multiple data links.
4. Transport Layer—Provides reliable transfer of data between end points, including end-to-end error recovery and flow control.
5. Session Layer—Responsible for establishing, maintaining, and terminating connections between communicating applications.
6. Presentation Layer—Allows independence to applications from differences in data representation at the end-user terminal.
7. Application Layer—The layer most visible to the end user, consisting of applications that perform information processing.

ATM is a technology that spans the Physical and Data Link layers.

2. B-ISDN Protocol Reference Model (PRM)

The B-ISDN protocol reference model is defined in ITU-T Recommendation I.321 and is shown in Figure 3-1 [2]. The B-ISDN PRM is a three-dimensional model consisting of planes in the vertical dimension:

Chapter 3: ATM Communications Protocols

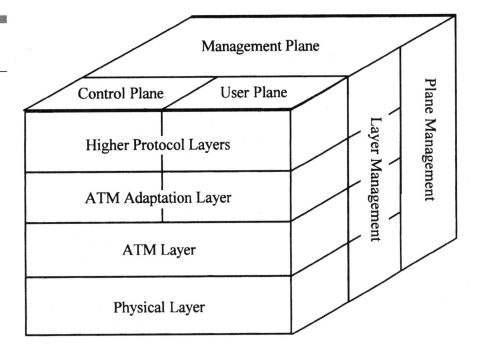

Figure 3-1.
B-ISDN protocol reference model.

1. Physical Layer—Corresponds to the Physical Layer in the OSI reference model.
2. ATM Layer—This layer, which corresponds to the Data Link Layer in the OSI reference model, is responsible for the transparent transfer of fixed-size ATM cells between communicating higher layer entities.
3. ATM Adaptation Layer (AAL)—Supports functions required by higher layer protocols, including mapping ATM cells to data packet formats used by those higher layer entities. While a Data Link layer protocol is generally concerned with communications over a single physical link between network nodes, the AAL performs its functions at the ends of an ATM connection, which may span multiple physical links. This gives the AAL a characteristic commonly associated with the Network Layer. Nevertheless, the primary functions performed at the AAL are Data Link functions: error detection, flow control, etc.
4. Higher Layers—Incorporates some (but not necessarily all) of the functionality of OSI RM Layers 3 through 7.

At the top face of the cube shown in Figure 3-1 is a second set of planes:

1. User Plane (U-plane)—As shown in Figure 3-1, the U-plane represents the set of functions that provide for the transfer of

information between two communicating end points. The U-plane contains a Physical Layer, ATM Layer, and an ATM Adaptation Layer that performs the function required by the Higher Layers. In relation to the OSI RM, the Higher Layers of the U-plane represent OSI RM Layers 3 through 7. Communications supported by the U-plane operate within the context of connections that have been previously established. Functions performed within the U-plane play no role in the establishment and termination of connections.

2. Control Plane (C-plane)—Connections between end points can be either permanent, or established on-demand. Functions performed in the C-plane provide for the on-demand establishment, and termination of connections between communicating end points. Higher Layers in the C-plane represent Network Layer functionality (such as routing) not performed by the Adaptation Layer.

3. Management Plane (M-plane)—The M-plane is concerned with management of the C- and U-planes, including error monitoring and reporting, connection verification, and performance monitoring. The M-plane also provides for the establishment of permanent connections between communicating end points. The M-plane consists of two types of functions:
   - Layer Management: performs layer-specific management functions (e.g., monitoring of bit error rates on a physical communications medium).
   - Plane Management: performs management and coordination functions that span across layers (e.g., establishment of a permanent connection).

The discussion in this chapter will focus on the U-plane. The operation of the C-plane will be discussed in the chapter on ATM Signalling, while M-plane operation will be discussed in the chapter on Operations, Administration, and Maintenance.

2.1 B-ISDN Protocol Operation

Figure 3-2 presents a conceptual depiction of the B-ISDN Protocol. Since we are focusing on the portion of the protocol that involves ATM specifically, higher layers are not presented. The same conceptual notions, however, extend to the higher layer protocols.

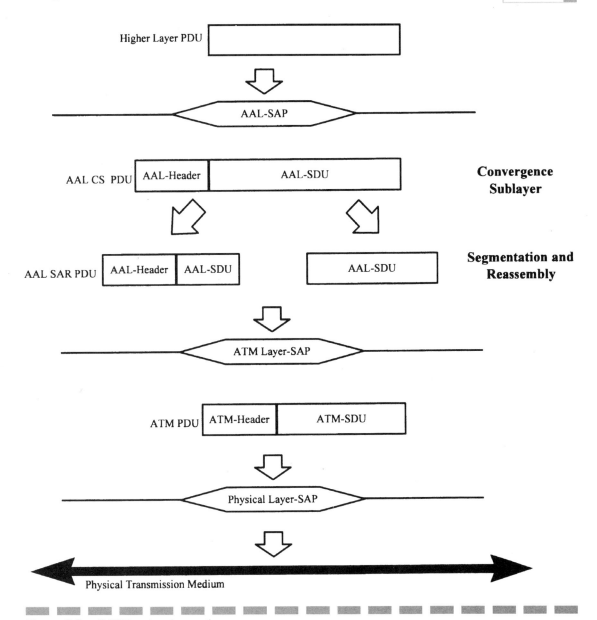

Figure 3-2. B-SISN protocol operation.

In Figure 3-2, the Higher Layer Protocol sends a packet of data, or Protocol Data Unit (Higher Layer PDU), to the AAL by accessing a Service Access Point (SAP) for the AAL (the AAL-SAP). The AAL-SAP defines a set of functions and data structures (or, "primitives") that are passed between the AAL and the Higher Layer. The SAP is uniquely addressed such that it corresponds to a specific connection between communicating Higher Layer entities. There may be one or more n-SAPs at the boundary between the Layer-n, and Layer-(n + 1) protocols.

At the AAL, the Higher Layer PDU becomes an AAL Service Data Unit (AAL-SDU) to which is appended an AAL-Header (and/or Trailer) containing information to support the functions performed by the AAL. Together the AAL-Header and/or Trailer, and AAL-SDU form the AAL-PDU. The maximum size of the AAL-PDU, which may be of variable length, is determined at the time that a connection is established between communicating endpoints.

In a generic layered protocol discussion, this AAL-PDU is passed as a unit to the next lower protocol layer which, in turn, appends its own header. The ATM layer accepts PDUs of a fixed size of 48 octets. Thus, Figure 3-2 shows that the AAL divides the AAL-PDU into smaller PDUs, which are then passed to the ATM Layer across the ATM Layer-SAP. Thus, the AAL Layer consists of two sublayers: a Convergence Sublayer (CS), which processes the Higher Layer PDU and appends an AAL Header; and a Segmentation and Reassembly (SAR), which divides the AAL-PDU into 48-octet SAR-PDUs at the originating end and, on the receiving end, reassembles the 48-octet units into the original AAL-PDU.

At the ATM Layer, a fixed 5-octet header is appended to SAR-SDU received across the ATM-SAP. This forms a 53-octet ATM-PDU commonly referred to as a "cell."

At the Physical Layer the cell is encoded into a bit stream for transport across a physical transmission medium.

3. The Physical Layer

The Physical Layer is defined in ITU-T Recommendation I.432, and in the ATM User-Network Interface Specification published by the ATM Forum [3][4]. At the time of this writing the ATM Forum had defined physical layer ATM interfaces for transmission rates ranging from 25.6 Mb/s to 622.08 Mb/s. The physical layer is functionally divided into two sublayers:

the Physical Media Dependent (PMD) Sublayer, and the Transmission Convergence (TC) Sublayer.

3.1 Physical Media Dependent (PMD) Sublayer

The PMD sublayer deals with characteristics, which are dependent upon the physical medium used. These include:

- Physical Medium Type: e.g., Category 5 Universal Twisted Pair, Single Mode Optical Fiber, etc.
- Bit Rate and Symmetry: the speed at which bits are carried over the medium and whether the bit rate is the same in both transmission directions.
- Maximum Length of Physical Medium: depends upon the signal attenuation characteristics of the physical medium used.
- Line Coding Method: e.g., Coded Mark Inversion, Non Return to Zero, etc.

3.2 Transmission Convergence (TC) Sublayer

The TC sublayer addresses physical layer characteristics, which are independent of the physical medium used. The TC is dependent upon the transmission format used for sending information across the physical medium. Regardless of transmission format, certain characteristics are common to all formats:

- Cell Delineation: the ability to distinguish individual ATM cells in the transmission frame—the method for accomplishing this, however, may vary for different transmission formats.
- Cell Error Detection: the ability to detect, and possibly correct, bit errors in the ATM cell.

Other TC sublayer characteristics are specific to the transmission format used over the physical medium. Thus, as shown in Table 3-1, the TC sublayer functions for SONET are, for the most part, different from those of DS-3 [5][6].

TABLE 3-1.

Transmission Convergence Sublayer Functions

SONET	DS-3
Cell Error Detection	Cell Error Detection
Cell Delineation[1]	Cell Delineation[2]
Cell Scrambling/Descrambling	Cell Scrambling/Descrambling
Path Signal Identification (C2 Octet)	PLCP Framing
Frequency Justification/Pointer Processing	Path Overhead Processing
Multiplexing	Clock Recovery
SONET Payload Scrambling/Descrambling	Nibble Stuffing
Transmission Frame Generation/Recovery	

Notes:
1. Cell Boundaries identified using HEC field in ATM Cell Header.
2. Cell Boundaries identified using PLCP Framing Octets

3.3 Physical Layer SAP

The Physical Layer-SAP is defined by:

- A PHY-UNITDATA.request primitive which is used by the ATM Layer on the sending end of a data link to inform the Physical Layer that an ATM cell is to be sent over the physical medium. The data structure for this primitive contains the 53-octet cell that is being sent.
- On the receiving end of a data link PHY-UNITDATA.indication informs the ATM Layer that an ATM cell has been received over the physical medium. The data structure for this primitive contains the 53-octet cell that is being received.

The address of the Physical-SAP is an index that identifies the specific physical interface port on the ATM switch to which the physical medium is attached.

4. The ATM Layer

4.1 ATM Cell Structure

The ATM Layer defines the structure of the fixed length ATM cell. It is specified in ITU-T recommendation I.361[7]. The format of the 53-octet

ATM cell at the UNI, shown in Figure 3-3, consists of a 5-octet cell header and a 48-octet payload. The ATM header contains the following fields:

- Generic Flow Control (GFC): This 4-bit field is of significance at the UNI only and is not carried end-to-end through ATM networks. At the NNI this field is absent. Instead, the bits in this position are included in the 12-bit VPI field.
- Virtual Path/Virtual Channel Identifier (VPI/VCI): The routing field of the ATM cell. Individual connections are uniquely identified by specific VPI/VCI assignments at the time the connection was established. The VPI field is 8 bits long at the UNI and 12 bits at the NNI. The VCI field is 16 bits. Some VPI/VCI values are reserved for special functions and are not available for user connections. Tables 3-2 and 3-3 show the reserved VPI/VCI values at the UNI and NNI respectively:
    - Idle cells are used for cell rate decoupling. Cell rate decoupling allows "filler" cells to be sent to make up for the difference between the rate at which cell traffic is transmitted by a sending entity and the physical layer transmission rate.

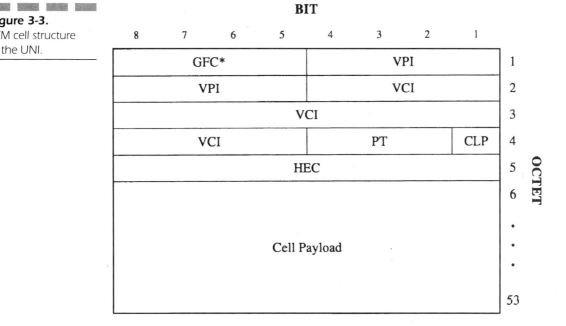

Figure 3-3.
ATM cell structure at the UNI.

TABLE 3-2.

Preassigned VPI and VCI Values at UNI

Use	VPI	VCI
Idle Cell	0	0
Invalid Cell	VPI ≠ 0	0
Meta-signalling	Any VPI value	1
General Broadcast Signalling	Any VPI value	2
Point-to-Point Signalling	Any VPI value	5
Segment OAM F4 Flow Cell	Any VPI value	3
End-to-End OAM F4 Flow Cell	Any VPI value	4
VP Resource Management Cell	Any VPI value	6
Reserved for future VP functions	Any VPI value	7
Reserved for other future functions	Any VPI value	32 > VCI > 7
Segment OAM F5 Flow Cell	Any VPI value	VCI ≠ 0, 3, 4, 6, or 7
End-to-End OAM F5 Flow Cell	Any VPI value	VCI ≠ 0, 3, 4, 6, or 7
VC Resource Management Cell	Any VPI value	VCI ≠ 0, 3, 4, 6, or 7
Reserved for future VC functions	Any VPI value	VCI ≠ 0, 3, 4, 6, or 7

TABLE 3-3.

Preassigned VPI and VCI Values at NNI

Use	VPI	VCI
Idle Cell	0	0
Invalid Cell	VPI ≠ 0	0
NNI Signalling	Any VPI value	5
Segment OAM F4 Flow Cell	Any VPI value	3
End-to-End OAM F4 Flow Cell	Any VPI value	4
VP Resource Management Cell	Any VPI value	6
Reserved for future VP functions	Any VPI value	7
Reserved for other future functions	Any VPI value	32 > VCI > 7
Segment OAM F5 Flow Cell	Any VPI value	VCI ≠ 0
End-to-End OAM F5 Flow Cell	Any VPI value	VCI ≠ 0
VC Resource Management Cell	Any VPI value	VCI ≠ 0, or 6
Reserved for future VC functions	Any VPI value	VCI ≠ 0

- The VPI/VCI reserved for general broadcast signalling is used in conjunction with meta-signalling. Note that meta-signalling is used at the UNI only.
- Resource management cells are used in conjunction with functions pertaining to traffic management.

■ Payload Type Identifier (PTI): PT is a 3-bit field which provides information about the cell payload to the receiving entity: whether it contains information being sent end-to-end between communicating entities (U-plane information), or whether the cell contains information to support M-plane functions. Table 3-4 shows values of the PTI field.
  - Congestion indication allows a receiving entity to determine if congestion was encountered while transporting the cell across the network. This information can be used at the connection endpoints to control the rate at which traffic is transmitted by performing traffic shaping (this topic will be discussed further in the chapter on traffic management). This congestion indication is referred to as the Explicit Forward Congestion Indication (EFCI). The EFCI is typically evaluated at each switching node along an end-to-end path. The EFCI bit is set to value "1" if any switch along the path was congested at the time that it processed the cell.

TABLE 3-4. Payload Type Identification

PTI Value	Interpretation
000	User data cell, congestion not experienced ATM user to ATM user indication = 0
001	User data cell, congestion not experienced ATM user to ATM user indication = 1
010	User data cell, congestion experienced ATM user to ATM user indication = 0
011	User data cell, congestion experienced ATM user to ATM user indication = 1
100	OAM F5 segment associated cell
101	OAM F5 end-to-end associated cell
110	Resource management cell
111	Reserved for future VC functions

- The ATM user to ATM user (AUU) field allows a receiving AAL entity to determine if more cells are required to reconstruct the AAL-PDU.
- Cell Loss Priority (CLP): This single bit field distinguishes cells with high priority of delivery from those with low delivery priority. Cells identified with low CLP are the first to be discarded if network conditions call for the discarding of ATM cell traffic. CLP is assigned to cells transported over a specific ATM connection based upon the Quality of Service class for that connection.
- Header Error Control (HEC): The HEC field is used by the Physical Layer to detect, and in some cases correct, bit errors in the ATM cell header. This field may also be used by the Physical Layer for cell delineation.

4.2 ATM Layer SAP

The ATM layer uses the VPI, VCI, and PTI values in the header to determine whether the cell is to be passed to the U-plane, C-plane, or M-plane. The Adaptation Layer entities for the U-plane and the C-plane use the same ATM layer SAP. The ATM layer exchanges information with an M-plane entity at the ATM layer and therefore does not use an SAP. In the U/C-planes, the ATM Layer-SAP consists of:

- ATM-DATA.request primitive: used by the AAL entity on the sending end of a connection to inform the ATM Layer that there is an ATM-SDU to be transported. This data structure contains the 48-octet ATM-SDU, and a CLP parameter whose value is inserted in the CLP field of the ATM cell header.
- When used by the ATM Layer entity in the receiving end of a connection, the ATM-DATA.indication primitive informs the AAL entity that an ATM-SDU has been received. This data structure contains the 48-octet cell that was received. This primitive also contains parameters that report information extracted from the ATM cell header: the CLP value of the received cell, and the congestion indication value.

The address of the ATM-SAP is an index that identifies a specific connection by the physical interface on the ATM switch and VPI/VCI.

5. ATM Adaptation Layer (AAL)

The Adaptation Layer occurs at the ends of a Virtual Channel Connection (VCC), which is the "end" of the connection as far as ATM is concerned. Thus, the Adaptation Layer is of an end-to-end nature that is more characteristic of a Network Layer protocol. However, unlike a Network Layer protocol, the AAL does not perform routing or connection establishment and termination; its primary functions involve checking the integrity of data sent across the network.

The AAL is an end-to-end Data Link protocol, which is key to the ability of ATM to provide end-to-end transport at high bit rates because it simplifies error checking at intermediate network nodes. Because of the AAL, the ATM Layer can limit error checking to a check of the 5-octet ATM header only. This task, which can be performed quickly based upon the first 5 octets of the received cell without the delay incurred while waiting for the rest of the cell to arrive, ensures only that the ATM network will deliver the cell payload to the right destination. Checking the integrity of information in the cell payload itself is performed only at the VCC endpoints by the Adaptation Layer.

ITU recommendation I.362 specifies a series of service classes that are the basis for AAL definition [8][1]. As shown in Table 3-5, the service classes are defined according to the type of traffic being carried:

The need to support these service classes led to the crafting of multiple AAL types. The functions and procedures performed by the AAL types are defined in ITU recommendations I.363, I.363.2, and I.366.1 [9][10][11]:

- AAL Type 1: This AAL supports services that transmit information at a constant bit rate. In addition, AAL 1 allows synchronous timing information to be exchanged between communicating entities across an ATM network.

TABLE 3-5. AAL Service Classifications

	Class A	Class B	Class C	Class D
Timing relation between source and destination	Required	Required	Not Required	Not Required
Bit rate	Constant	Variable	Variable	Variable
Connection mode	Connection-oriented	Connection-oriented	Connection-oriented	Connectionless

---

[1] I362 has been withdrawn as a formal recommendation but is mentioned here for historical context.

- AAL Type 2: AAL 2 provides for bandwidth-efficient transmission of short (less than 48 octets) packets over ATM networks. Such packets may be used in low bit rate, delay-sensitive applications. This includes variable bit rate audio and video.
- AAL Type 3/4: the result of a merging of the definitions of AAL Type 3, and AAL Type 4: AAL Type 3 being defined for connection-oriented data services, and AAL Type 4 providing connectionless data services.
- AAL Type 5: A simplified version of AAL 3/4 that provides support for connection-oriented data services.

A fifth type of AAL is the "null" AAL, referred to as AAL Type 0. This AAL type represents the case where no AAL function is performed and ATM cells are passed directly to the higher layer entity for processing.

The functions performed by a given AAL type may differ according to the higher layer protocol being supported. Thus, the AAL definitions are structured as some combination of the following sublayers:

- Segmentation and Reassembly (SAR): as described earlier, this is the procedure that segments a variable length AAL PDU into fixed length ATM cells on the originating end of the VCC, and reassembles cells into variable length PDUs at the terminating end of the connection. While the end result is the same for all AAL types, the structure of the ATM cell payload differs among AAL types.
- Common Part Convergence Sublayer (CPCS): performs error checking, extraction of synchronous timing information, reassembly of long-length PDUs, and other functions on the PDU prior to segmentation and following reassembly. CPCS functions differ for different AAL types but are fixed for a given AAL type.
- Service Specific Convergence Sublayer (SSCS): this sublayer most directly supports the specific functions needed by a given higher layer protocol. As an example, the SSCS may be required to inspect Quality of Service (QoS) information in the header of the higher layer SDU (which becomes the AAL SDU when passed across the AAL-SAP), and for passing that information to the ATM Layer to ensure that the ATM connection provides similar QoS, as is expected by the higher layer protocol. For a given AAL type, the SSCS represents a framework for defining the specific functions to be performed for a given higher layer protocol. Thus the SSCS must have procedures for determining how to extract information from the higher layer PDU to identify to specific protocol being handled.

G.711, Frame Relay, or SMDS, and TCP/IP are examples of higher layer protocols that may be transported over ATM networks. SSCS specifications are generally stated in separate recommendations specific to the higher layer protocol being served.

5.1 AAL Type 1

AAL 1 provides two types of service:

1. Unstructured Data Transfer service, in which data are treated as a continuous octet stream.
2. Structured Data Transfer service, in which data are grouped into identifiable blocks of arbitrary size.

In either case, the length of the AAL-SDU must be constant, as must also be the time interval between SDU arrivals across the SAP.

5.1.1 SAR Sublayer

The basic SAR-PDU is shown in Figure 3-4. The 48-octet SAR-PDU consists of a 47-octet payload, and a single header octet containing two fields:

- Sequence Number (SN): consisting of a Convergence Sublayer Indication (CSI) bit, and a 3-bit Sequence Count Field. Values for each of these subfields is set by the Convergence Sublayer in the sending AAL entity. The Sequence Count field is a modulo 8 counter that allows the receiving AAL entity to detect loss of SAR-PDUs. The CSI can be used for multiple purposes. The communicating AAL entities negotiate the significance of this bit at connection establishment.
- Sequence Number Protection (SNP): provides a mechanism for detecting and/or correcting errors in the SAR-PDU header.

5.1.2 Convergence Sublayer

The CS of AAL 1 provides a number of procedures for use in the transport of constant bit rate (CBR) sources. The use of these procedures,

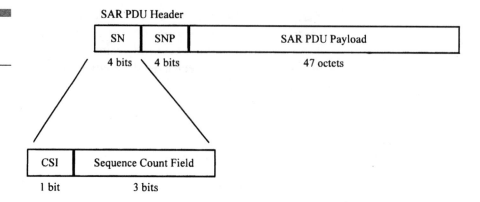

Figure 3-4.
SAR-PDU format for AAL 1.

however, depends upon the type of service (circuit, video, voice-band, or high quality audio) being carried. Likewise, the size of the AAL-SDU also depends upon service type. Examples of services using AAL 1 are discussed in Chapter 7 ("ATM Service Interworking").

In general, the CS consists of the following procedures:

1. Sequence Count Operations

2. Source Clock Frequency Recovery Method

3. Structured Data Transfer Method

5.1.2.1 Sequence Count Operations In the transmitting end, the CS assigns a value to be used in the sequence count field of the SAR-PDU. This value begins with 0 and is incremented sequentially modulo 8. On the receiving end the sequence count field is checked.

5.1.2.2 Source Clock Recovery Method The two methods used for source clock recovery by AAL 1 are:

- Synchronous Residual Time Stamp (SRTS)
- Adaptive Clock Method

The SRTS method allows a receiver to reconstruct the source clock based upon a common network clock, and the frequency difference between the source clock and the common network clock. In SRTS the CS sends a 4-bit encoding of this difference, called a Residual Time Stamp (RTS), bit-serially in the CSI field of odd-numbered (based upon the sequence count field) SAR-PDUs. There is not a standardized method to handle the case in which there is not a common network clock between the communicating entities at the VCC endpoints.

In the Adaptive Clock Method, the receiver maintains a buffer that is read out under control of a local clock. By monitoring the fill level of the buffer (above or below the median buffer level), the local clock rate is increased or decreased to prevent buffer overflow or underflow.

5.1.2.3 Structured Data Transfer (SDT) Method The CS for AAL 1 uses this method for the transport and reassembly of AAL-PDUs that cross ATM cell boundaries. There is no size constraint on the size of the block, only that it be of a fixed, octet-based structure. The bounds of the structure are defined by a pointer carried in the SAR-PDU. This results in two formats for the SAR-PDU:

- Non-P (pointer) Format: Identical to the SAR-PDU shown in Figure 3-4 with the constraint that CSI-0 in even (segment count) numbered SAR-PDUs. Odd-numbered SAR-PDUs are always Non-P format.
- P Format: In this format, the first octet of the SAR-PDU is a Pointer Field as shown in Figure 3-5. P-Format SAR-PDUs may only occur in even numbered sequence count values, and are indicated by CSI-1.

The value of the pointer field indicates the offset (in octets) between the end of the pointer field and the start of the first octet of the next structure block (which, by implication, identifies the end of the current structure). Note that the SDT method uses the CSI in such a manner so as to not conflict with the SRTS method. Since P format SAR-PDUs may only occur on every other ATM cell at most, the pointer value spans the range of values from 0 to 93. The value 93 is special in that it indicates that a structure block ends at the end of the next cell, whose beginning is not within either the current or next cell.

5.1.3 AAL 1 SAP

The AAL 1 SAP consists of send and received data primitives for each of the services that it provides:

Figure 3-5.
P Format SAR-PDU.

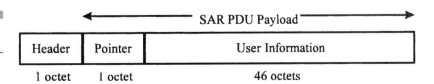

- Unstructured Data Transfer Service:
  - AAL-UNITDATA-REQUEST primitive used by the Higher Layer Protocol on the sending end of a connection to inform the AAL Layer that there is an AAL-SDU to be transported.
  - AAL-UNITDATA-INDICATION primitive used by the AAL Layer on the receiving end of a connection to inform the Higher Layer Protocol that an AAL-SDU has been received. This data structure contains the data that was received.
- Structured Data Transfer Service:
  - AAL-UNITDATA-REQUEST primitive in which the Higher Layer Protocol also passes a structure indication of whether the AAL-SDU marks the start, or continuation, of a structure block.
  - AAL-UNITDATA-INDICATION primitive containing the data that was received, the structure indication, and an indication of whether the data received is judged to be errored or nonerrored.

Each AAL 1 connection is associated with a single ATM Layer connection. Therefore, the AAL 1 SAP address is a connection identifier that has a one-to-one mapping to an ATM Layer connection identifier.

5.2 AAL Type 2

AAL 2 is subdivided into a Common Part Convergence Sublayer, and three Service Specific Convergence Sublayers as shown in Figure 3-6. As we can see in the figure, AAL 2 presents three AAL-SAPs to the higher layer protocol—one for each SSCS:

1. Service Specific Segmentation and Reassembly Sublayer (SSSAR), which represents "basic" SSCF functionality
2. Service Specific Transmission Error Detection Sublayer (SSTED)—provides enhanced error detection in addition to basic SSSAR services.
3. Service Specific Assured Data Transport Sublayer (SSADT)—uses SSTED error detection capabilities and adds reliable data transport services.

As shown in Figure 3-6, the 3 SSCFs form a hierarchy with the "higher" SSCF enhancing the services provided by the "lower" SSCF.

Chapter 3: ATM Communications Protocols

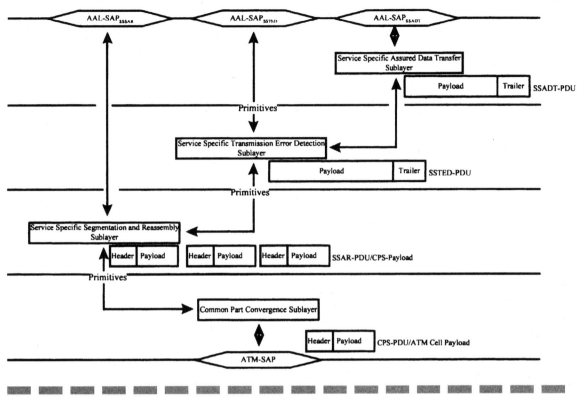

Figure 3-6. AAL type 2 structure.

5.2.1 Common Part Sublayer

The Common Part Sublayer (CPS) of AAL 2 on the sending side receives a variable length CPS-Packet and attempts to pack it, along with other CPS-Packets into a single ATM cell payload (the process is reversed on the receiving side). As shown in Figure 3-7, if we have CPS-Packets from multiple VCCs, each VCC must have its own instance of an SSCS uniquely identified by a Connection Identifier (CID) that will allow the CPS to distinguish which portions of an ATM cell payload are associated with each individual VCC. Note that the ATM Layer is unable to distinguish individual connections within an ATM Payload.

The format of the CPS-Packet is shown in Figure 3-8. It is comprised of a CPS-Packet Payload and a Header. The Header field consists of:

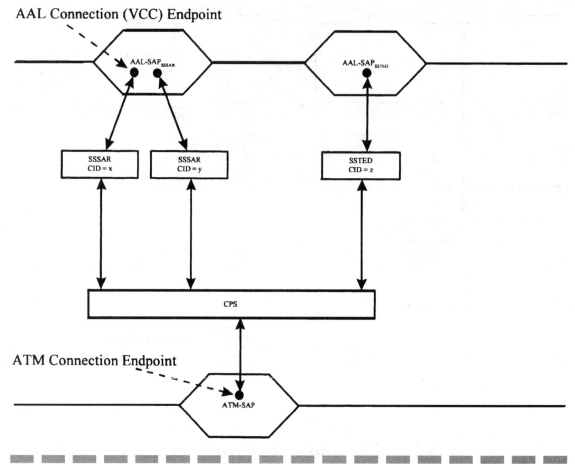

Figure 3-7. VCC multiplexing in AAL 2.

- Connection Identifier (CID)
- Length Indicator (LI), which defines the size (in octets) of the CPS-Packet Payload
- User-to-User Indication (UUI)—carried transparently end-to-end between SSCS entities
- Header Error Control (HEC)—provides error detection, and possibly correction of errors in the CPS-Packet Header

Assembly of individual CPS-Packets results in the 48-octet CPS-PDU of Figure 3-9:

- Offset Field (OSF), which gives the offset location (in octets) from the beginning of the CPS-PDU Payload, to the beginning of the first CPS-Packet. Once the first CPS-Packet in the CPS-PDU Payload has been located, LI fields may be used to locate any other CPS-Packets in the CPS-PDU Payload. If no CPS-Packet starts in this CPS-PDU Payload, the OSF either indicates the beginning of the PAD field (which implicitly gives the location of the end of a CPS-Packet), or indicates that there is no start boundary in the CPS-PDU Payload.
- Sequence Number (SN)—a single bit field which can provide an indication of out-of-sequence delivery if a single (or odd number) CPS-PDU is lost in transmission.
- Parity—provides error detection over the Start Field (which consists of the OSF, SN, and Parity fields)
- CPS-Packet—a field containing a variable number of CPS Packets and/or partial packets
- PAD—fills out CPS-PDU Payload to 47 octets in the event that there is insufficient data in CPS-Packet(s)

The interface between the CPCS and SSCS is defined by the following primitives:

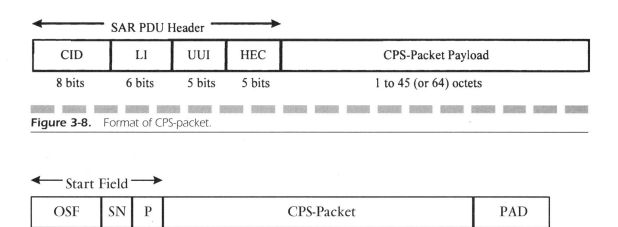

Figure 3-8. Format of CPS-packet.

Figure 3-9. CPS-PDU.

- CPS-UNITDATA.request primitive used by the SSCS to inform the CPCS that there is data to be transported. The data structure contains a CPS-Packet Payload, and a user-to-user information parameter.
- CPS-UNITDATA.indication primitive is used by the CPCS to inform the SSCS of the arrival of a CPS-PDU.

5.2.2 Service Specific Convergence Sublayers

5.2.2.1 Service Specific Segmentation and Reassembly Sublayer (SSSAR) On the transmit side of a connection, the SSSAR receives an SDU either from a higher layer protocol directly, or indirectly through the SSTED. The SSSAR may then segment this SDU (which can be up to 65,568 octets in length) into CPS-Packets. The maximum size of a CPS-Packet may be specified, on a per-VCC basis, as either 45 or 64 octets. These CPS-Packets are then sent to the AAL 2 Common Part Sublayer along with the User-to-User field populated with a value to indicate whether more CPS-Packets are required to reconstruct the SDU at the receiving end.

5.2.2.2 Service Specific Transmission Error Detection Sublayer (SSTED) The SSTED adds fields to the Trailer of an SDU (from the Higher Layer protocol, or the SSADT Sublayer) as shown in Figure 3-10. The fields in the SSTED-PDU Trailer provide the peer AAL entity on the receiving side of the connection with enhanced ability to detect errors. The Cyclical Redundancy Check (CRC) field detects bit errors in the SSTED-PDU, while the Length field provides a mechanism to determine the expected length (in octets) of the SSTED-PDU Payload. Disparities between the actual and expected Payload length indicate that the Payload has been corrupted by the loss of information and/or merging of information from other SSTED-PDUs. Other fields in the SSTED-PDU Trailer are CLP, Congestion Indication (CI), and User-to-User (SSTED-UU). The CLP and CI fields are not processed by the SSTED Sublayer, but are provided for compatibility with the AAL 5SAP. The User-to-User field is also carried transparently by the SSTED Sublayer.

5.2.2.3 Service Specific Assured Data Transfer Sublayer (SSADT) SSADT is based upon the assured data transfer method of the Service Specific Connection Oriented Protocol (SSCOP), which is defined in ITU recommendation Q2110 [12]. The basic principle of reliable data transport is that PDUs from a sender carry a sequence number

Chapter 3: ATM Communications Protocols

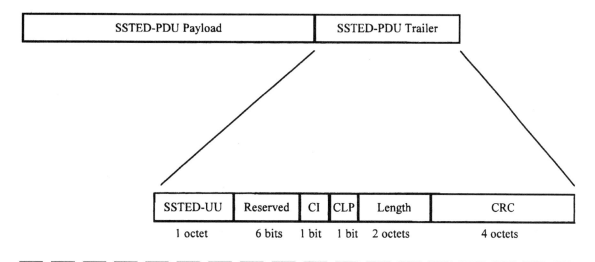

Figure 3-10. SSTED-PDU.

(Sequenced Data PDU) which allows the receiver to detect missing PDUs. Being a connection-oriented (as opposed to connectionless) network service, ATM guarantees that PDUs will be sent across the network in sequence. Therefore, an out-of-sequence PDU indicates that data has been lost in transport. Use of a large sequence count modulus (modulo 2^{24}) effectively "guarantees" that lost PDUs will be detected.

When the sender detects that PDU loss has occurred, it selectively retransmits only those PDUs that have been detected as lost. The mechanism for detecting PDU loss is through a poll and response method that works as follows:

1. At regular intervals, the sender will query the receiver (via a Poll PDU) to determine the sequence number of the last received PDU.

2. In response to the query from the sender, the receiver will send a response (via a Solicited Status PDU) to inform the sender which PDUs have been received.

3. If the response indicates that one or more PDUs have been lost in transmission, the sender will resend only those PDUs as indicated in the response. This requires that the sender must store a copy of all sent PDUs until receiving acknowledgment of receipt from the receiver.

4. If the receiver, however, receives a PDU whose sequence number is out of sequence, it is assumed that PDU loss has occurred. The receiver will then autonomously send an Unsolicited Status PDU

to indicate which PDUs have been lost. Any PDUs received out of sequence are held in a buffer pending receipt of the missing PDUs.

5. In response to an Unsolicited Status PDU from the receiver, the sender will selectively retransmit any lost PDUs.

6. Once the missing PDUs are received, the associated SDUs, along with any previously buffered (out-of-sequence) SDUs, are delivered to the Higher Layer entity.

The Sequenced Data, Poll, Solicited Status, and Unsolicited Status PDUs are shown in Figures 3-11, 3-12, 3-13, and 3-14. All PDUs use PAD fields, to ensure that they are aligned to 4 octet boundaries.

In the Sequenced Data PDU, the Data Sequence Number field contains the modulo 2^{24} sequence number of the PDU.

In the Poll PDU:

- *Sender Poll SN* is the sequence number of the Poll PDU
- *Sender Next Expected Data SN* is the expected sequence number of the next Sequenced Data PDU from the sender

In the Solicited Status PDU:

- *First Missing PDU SN* (if present) is the lowest PDU SN that is detected as missing by the receiver.

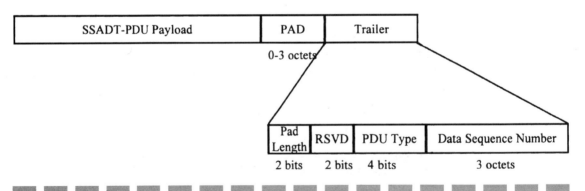

Figure 3-11. Sequenced data PDU.

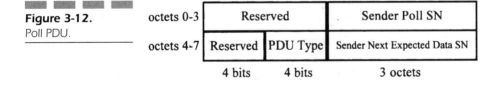

Figure 3-12. Poll PDU.

Chapter 3: ATM Communications Protocols

Figure 3-13.
Solicited status PDU.

octets 0-3	PAD	First Missing PDU SN
octets 4-7	PAD	Next Received PDU SN
octets 8-11	PAD	Second Missing PDU SN
octets 12-15	PAD	Next Higher Received PDU SN
	⋮	
octets N - N+3	PAD	Next Highest Expected PDU SN
octets N+4 - N+7	Reserved	Received Poll SN
octets N+8 - N+11	Reserved	Receiver Maximum SN
octets N+12 - N+15	Reserved | PDU Type	Receiver Next Expected Data SN
	4 bits 4 bits	3 octets

Figure 3-14.
Unsolicited status PDU.

octets 0-3	PAD	First Missing PDU SN
octets 4-7	PAD	Next Received PDU SN
octets 8-11	Reserved	Receiver Maximum SN
octets 12-15	Reserved | PDU Type	Receiver Next Expected Data SN
	4 bits 4 bits	3 octets

- *Next Received PDU SN* (if present) is the sequence number of the next received Sequenced Data PDU.
- *Second Missing PDU SN* (if present) indicates that there is a break in the PDU delivery sequence. This parameter reports the lowest PDN SN that is detected as missing whose SN is higher than that of the Next Received PDU SN.
- *Next Higher Received PDU SN* (if present) is the sequence number of the next received Sequenced Data PDU whose SN is higher than that of the Second Missing PDU SN.

The list continues listing pairs as above for all detected breaks in the PDU delivery sequence:

- *Next Highest Expected PDU SN* is one greater than the highest Sequenced Data PDU SN received. In the event of lost PDU(s) this value is set to the value in the Sender Next Expected Data SN field of the Poll PDU. This action prevents generation of Unsolicited Status PDUs for lost PDUs that fall within a range that have already been reported to the sender through a Solicited Status PDU.
- *Received Poll SN* is the sequence number of the Poll PDU from the sender for which this response is being generated.
- *Receiver Maximum SN* is the current maximum Sequenced Data PDU SN that the receiver will accept.
- *Receiver Next Expected Data SN* is the sequence number of the next Sequence Data PDU that is expected by the receiver. This is the same as the First Missing PDU SN.

The assured data transfer service also implements flow control to help prevent data loss as a result of a flood of data being transmitted from the sender at a rate faster than the receiver can accept. The mechanism used is that of a sliding window [1]:

1. The receiver informs the sender of the highest sequence numbered PDU that it will accept at the time of connection establishment.
2. The sender sequentially transmits PDUs to the sender as long as no lost PDUs are indicated by the receiver, and as long as the sequence number does not exceed the maximum.
3. As Solicited or Unsolicited Status PDUs are received from the receiver, the maximum sequence number is updated by the sender to reflect the value in the received Status PDU.
4. If the sender arrives at the maximum PDU sequence count before receiving an update from the receiver, transmission of PDUs from the sender is halted until the receiver authorizes a higher count.

Figures 3-15, 3-16, and 3-17 give examples of assured data transport using a window size of 4. Figure 3-15 shows the case in which operation is error-free and no PDUs are lost. After delivery of 4 PDUs, a Poll PDU is transmitted from the sender to the receiver. The Poll PDU contains what the sender believes to be the next sequence PDU expected by the receiver (PDU #5), and an indication that this Poll PDU has a sequence number of 1 (i.e., this is the first Poll PDU). The Solicited Status PDU sent in response by the receiver confirms that the receiver expects

PDU #5 next, that the response is to Poll PDU #1, and that the new maximum sequence count is 8.

In Figure 3-16 we see the case in which the second, third, and fourth transmitted PDUs are lost. In this case, the receiver doesn't know that there is anything wrong since no PDUs have arrived out of sequence. Only after the sender issues a Poll, and receives a Solicited Status response does the sender detect that PDUs have been lost. The sender then retransmits PDUs #2, #3, and #4. Through the Solicited Status response, the receiver indicated that the maximum PDU sequence count is now 5. This allows the sender to transmit PDU #5 after retransmitting PDUs #2, #3, and #4.

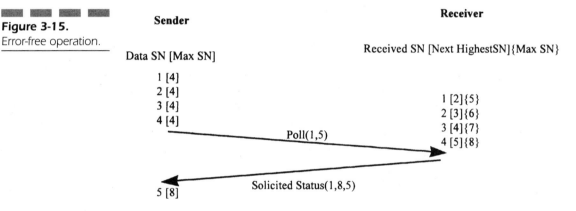

Figure 3-15.
Error-free operation.

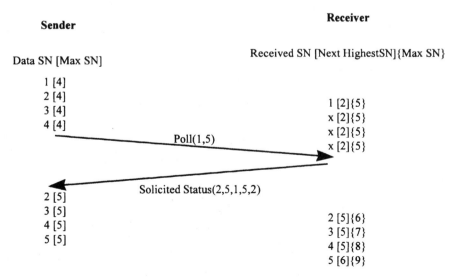

Figure 3-16.
Error recovery via solicited status PDU.

Figure 3-17.
Error recovery via unsolicited status PDU.

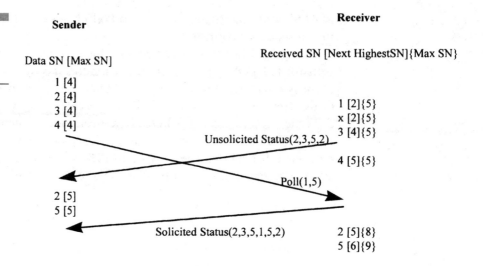

In Figure 3-17, PDU #2 is lost. Upon receipt of PDU #3, the receiver sends an Unsolicited Status PDU. Prior to receiving this PDU, the sender issues a Poll of the receiver and waits for a Status Response. The Unsolicited Status PDU arrives first, from which the sender can only determine that PDU #1 and PDU #3 have been received, and that PDU #2 was not successfully received. Note that loss of PDU #4 would be indicated by a separate Unsolicited Status PDU (and certainly would be indicated in the Solicited Status PDU response to the Poll PDU). After retransmitting PDU #2, the sender then transmits PDU #5.

In this last example, note that when PDU #2 is received, PDU #2, PDU #3, and PDU #4 may be delivered to the Higher Layer protocol. This allows the Maximum Sequence Number value to jump from 5 to 8 upon receipt of PDU #2 as shown in Figure 3-17.

5.2.3 AAL 2 SAP

AAL 2 provides 3 SAPs, each of which provides the Higher Layer protocol an interface to each individual Service Specific Convergence Sublayer:

- The SAP for SSSAR consists of:
  - SSSAR-UNITDATA.request primitive used by the Higher Layer protocol to inform the SSSAR that there is data to be transported. This data structure contains an SSSAR-SDU, and a user-to-user information parameter. This second parameter

provides a mechanism for the out-of-band exchange of messages between the Higher Layer entities at the VCC endpoints.
- SSSAR-UNITDATA.indication primitive used by the SSSAR to inform the Higher Layer protocol of the arrival of an SSSAR-PDU. The parameters are the same as those for the send data request primitive.

- The SAP for SSTED consists of:
  - SSTED-UNITDATA.request primitive used by the Higher Layer protocol to inform the SSTED that there is data to be transported. This data structure contains an SSTED-SDU, and a user-to-user information parameter. The SSTED data request primitive also has parameters that allow the sending entity to indicate values for CLP and CI parameters, which are transported transparently between VCC endpoints.
  - SSTED-UNITDATA.indication primitive used by the SSTED to inform the Higher Layer protocol of the arrival of an SSTED-PDU. The parameters are the same as those for the send data request primitive.

- The SAP for SSADT is as defined in the SSCOP, which is discussed in Chapter 4.

The AAL 2 SAP address is associated with a connection identifier, which uniquely identifies a VCC.

5.3 AAL Type 3/4

AAL 3/4 is the result of merging of two AAL definitions: AAL 3, which was intended to serve applications with connection-oriented, variable bit rate traffic; and AAL 4, which was specified to serve connectionless VBR applications. The result, AAL 3/4, serves both connection-oriented VBR applications and provides adaptation of the connectionless protocol to the intrinsically connection-oriented ATM layer protocol.

While it is a misnomer to state that AAL 3/4 has features in common with AAL 2 (since AAL 3/4 was defined earlier), both support multiplexing of traffic from different AAL connection endpoints into a single ATM connection. The equivalent to the CID of AAL 2, is the Multiplexing Identifier (MID) in AAL 3/4. One key difference, between the two adaptation layers, however, is that in AAL 3/4 a single ATM cell may only carry traffic from a single AAL connection endpoint. Traffic from different AAL connection endpoints may be interleaved within a single ATM cell stream.

5.3.1 SAR Sublayer

Figure 3-18 shows the AAL 3/4 SAR-PDU, which contains the following fields:

- Segment Type (ST)—identifies whether there are more SAR-PDUs required to reconstruct the SAR-SDU. The ST field identifies four types of segments:
    1. Beginning of Message (BOM), which indicates that the SAR-PDU contains the beginning of the SAR-PDU.
    2. Continuation of Message (COM).
    3. End of Message (EOM).
    4. Single Segment Message (SSM)—indicating that the SAR-PDU contains the entire contents of the SAR-SDU.
- Sequence Number (SN)—allows SAR-PDUs belonging to an SAR-SDU to be sequence-numbered modulo 16.
- Multiplexing Identifier (MID)—identifies SAR-SDUs belonging to a specific AAL connection endpoint. All SAR-PDUs associated with a single SAR-SDU will have the same MID value.
- SAR-PDU Payload—the payload is always 44 octets in length but may contain a PAD trailer if this is an EOM or SSM containing less than 44 octets of SAR-SDU information.
- Length Indication (LI)—indication of the number of SAR-SDU octets in the SAR-PDU.
- CRC—provides error detection over the SAR-PDU.

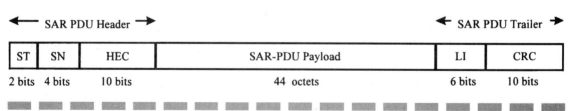

Figure 3-18. Format of SAR-PDU for AAL 3/4.

Chapter 3: ATM Communications Protocols

Note that if an EOM is lost in transmission, followed by a multiple of 16 complete SAR-SDUs and the following BOM, it is possible that SAR-PDUs from different SAR-SDUs can be reassembled into a single SAR-SDU by the SAR.

5.3.2 Convergence Sublayer

The CPCS for AAL 3/4 provides two types of service:

- Message Mode Service: the CPCS-SDU (as received from the SSCS, or across the AAL-SAP) is sent to the SAR sublayer in a single CPCS-PDU.
- Streaming Mode Service: the CPCS-SDU is sent to the SAR sublayer in multiple CPCS-PDUs. This service conceptually performs the function of the AAL 2 Service Specific Segmentation and Reassembly sublayer.

As shown in Figure 3-19, the CPCS-PDU contains:

- Common Part Indicator (CPI)—generally set to indicate that the octet is the unit of measure for length fields in the PDU header.
- Beginning Tag (Btag)—the sender generates a value for use in this field, along with the same value in the Etag field of the Trailer, for each CPCS-PDU. Comparing the Btag and Etag values allows the receiver to detect lost CPCS-PDU Headers and Trailers.
- Buffer Allocation Size (BASize)—an indication from the sender to inform the receiver of the amount of buffering required to receive the CPCS-SDU. The BAsize unit of measure is as defined in the CPI field. Note that for message mode, this is the same as the length of the CPCS-PDU Payload, while in streaming mode this is typically a value greater than the length of the Payload.
- CPCS-PDU Payload—is the CPCS-SDU in message mode, but a portion of the CPCS-SDU in streaming mode.
- PAD—may range from 0 to 3 octets in size to ensure 32-bit alignment of the CPCS-PDU Payload.

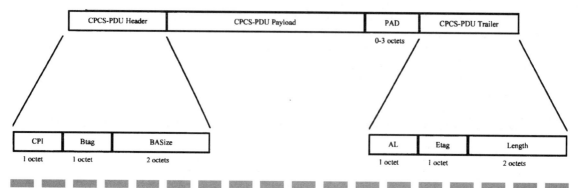

Figure 3-19. CPCS-PDU for AAL 3/4.

- Alignment (AL)—contains no information, the only purpose of field is to ensure 32-bit alignment of the Trailer.
- End Tag (Etag)—contains same value as in Btag field.
- Length—indicates length of the CPCS-PDU Payload field. The Length unit of measure is as defined in the CPI field.

5.3.3 Connectionless Network Service

The provision for broadband connectionless data service is defined in ITU recommendation I.364 [13]. The CPCS of AAL 3/4 has the basic functionality to support connectionless network layer protocols without the need for a service specific convergence sublayer.

Connectionless service is provided over ATM networks by means of a Connectionless Service Function (CLSF). The CLSF provides a Network Layer protocol, the Connectionless Network Access Protocol (CLNAP), that sits on top of AAL 3/4 as shown in Figure 3-20.

The function of the CLSF is to give the connection-oriented ATM network the appearance of being connectionless to higher layer protocols. This provides an overall end-to-end service in which a connectionless protocol sits on top of a connection-oriented protocol—the opposite of the case of TCP/IP (in which a connection-oriented protocol, TCP, sits on top of a connectionless protocol, IP) [14].

The CLSF is a logical entity that may exist in many locations throughout the network as shown in Figure 3-21. The CLSF may be located within a Private Connectionless Network, or within the network of a Specialized Provider of connectionless service, or within the public

B-ISDN network itself. Furthermore, the CLSF may be provided by an ATM switch, or provided externally by a module with a direct connection to an ATM switch, such as a router.

The CLSF provides functions such as routing, addressing, and QoS selection. Thus, when the CLSF receives a datagram from the connection-

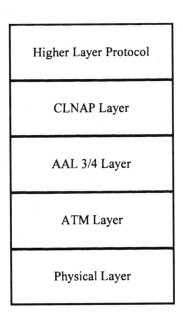

Figure 3-20.
Protocol structure for connectionless data service in B-ISDN.

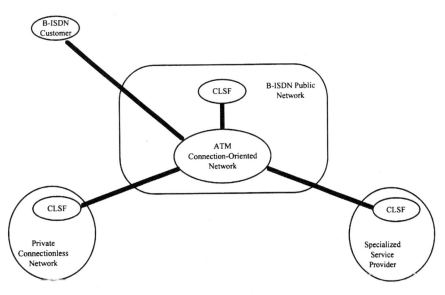

Figure 3-21.
B-ISDN connectionless service configuration.

less protocol, the connectionless server associates the destination endpoint (as specified in the Destination Address field within the datagram) with a route through the underlying ATM network. If a connection between the CLSF and the destination already exists, the routing function is a simple table lookup to match the Destination Address with an ATM (VPI/VCI) address. If no such connection exists, the CLSF must interact with the C-plane in the underlying ATM network to establish a connection.

In the example of a connectionless service configuration shown in Figure 3-22, the connectionless service between communicating endpoints involves two connections:

1. A connection between the source and the CLSF. While there is no need for the source to know how to reach the destination, it must know how to reach the CLSF. This connection may be either a PVC or SVC.

2. A connection between the CLSF and the destination. This connection may also be a PVC or SVC.

An example of connectionless service over AAL 3/4 using Switched Multimegabit Data Service is discussed in Chapter 7.

5.3.4 AAL 3/4 SAP

The SAP for AAL 3/4 consists of the following primitives:

- For Message Mode service:

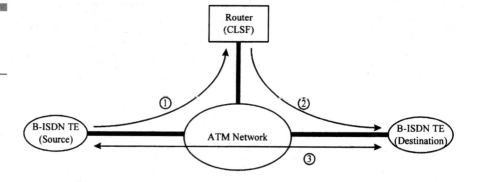

Figure 3-22. ATM connections for connectionless service.

1. Source to CLSF Connection
2. CLSF to Destination Connection
3. Source to Destination Connectionless Service

- CPCS-UNITDATA request primitive used by the sending Higher Layer protocol entity to inform the CPCS that there is data to be transported. This data structure contains an Interface Data Unit (IDU) from the Higher Layer protocol.
- CPCS-UNITDATA indication primitive used by the receiving CPCS entity to inform the Higher Layer protocol of the arrival of a CPCS-PDU. The parameters include those for the send data request primitive, in addition to a Reception Status parameter to inform the Higher Layer protocol whether the data being delivered is corrupted.

- For Streaming Mode service:
  - In streaming mode service, the CPCS-UNITDATA primitive used by the sending higher layer protocol entity contains additional parameters not used in message mode service: a More parameter to indicate whether there are additional IDUs required to form a complete CPCS-SDU, and a Maximum Length parameter to indicate the maximum length of a CPCS-SDU.
  - CPCS-UNITDATA primitive used by the receiving CPCS entity also contains the Reception Status parameter.

The MID associated with an AAL 3/4 connection endpoint is associated with the AAL 3/4 SAP address. Since the MID must be unique within an ATM Layer connection, the AAL 3/4 SAP address must provide an index to identify both a MID and an ATM Layer connection.

5.4 AAL Type 5

AAL 5 is the result of attempts to offer AAL 3/4 connection-oriented services for data applications, but with reduced overhead. Thus, AAL 5 was previously known as SEAL (Simple and Efficient AAL). As a result of simplifications, AAL 5 offers services similar to those of AAL 3/4 with the following exceptions:

- No support for multiplexing of traffic from different AAL connection endpoints into a single ATM connection. Thus, there is a one-to-one mapping between AAL connection endpoints and ATM layer connections.
- AAL 5 does not embed a Segment Type indication in the SAR-PDU as is done in AAL 3/4. Instead, AAL 5 uses a bit in the Payload Type field of the ATM Cell Header to identify the end of an SAR-SDU.

No distinction is made between a beginning of message and continuation of message.

- AAL 5 provides no error detection capabilities over the SAR-PDU. All error checking at the receiving end of a connection is performed over the reassembled CPCS-PDU.

AAL 5, like AAL 3/4, offers message mode and streaming mode services, in addition to assured and nonassured data transfer.

The Convergence Sublayer of AAL 5 consists of a CPCS and an SSCS. SSCS protocols are defined in separate ITU recommendations. Chapter 7 discusses an example using a Frame Relay SSCS over AAL 5.

5.4.1 SAR Sublayer

The SAR-PDU for AAL 5 consists entirely of Payload, with no overhead.

5.4.2 Convergence Sublayer

The CPCS-PDU for AAL 5 contains the following fields, as shown in Figure 3-23:

- CPCS-PDU Payload—is the CPCS-SDU in message mode, but a portion of the CPCS-SDU in streaming mode. This field may be up to 65,535 octets in length.
- PAD—may range from 0 to 47 octets in size to ensure that the CPCS-PDU is a multiple of 48 octets.

Figure 3-23. CPCS-PDU format for AAL 5.

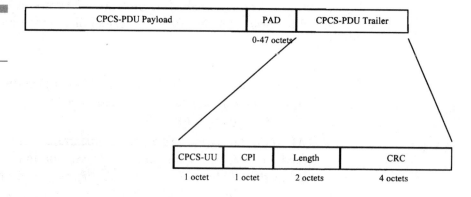

- CPCS User-to-User indication (CPCS-UU)—contains information which is transferred transparently between peer CPCS entities. This field allows Higher Layer protocols to exchange information out of band (i.e., outside of the CPCS-Payload).
- Common Part Indicator (CPI)—provides 8-octet alignment for the AAL 5 Trailer.
- Length—indicates length of the CPCS-PDU Payload field.
- Cyclical Redundancy Check (CRC)—used to detect bit errors in the CPCS-PDU.

5.4.3 AAL 5 SAP

The AAL-SAPs presented to Higher Layer protocols depends upon the Service Specific Convergence Sublayer protocol being used. The CPCS presents the following primitives to the SSCS:

- For Message Mode service:
  - AAL-UNITDATA.request primitive used by the SSCS to inform the CPCS that there is data to be transported. This data structure contains an Interface Data Unit (IDU) from the SSCS, a Loss Priority (CPCS-LP) parameter, a Congestion Indication (CPCS-CI) parameter, and a User-to-User indication (CPCS-UU). The CPCS-LP and CPCS-CI values are mapped into the ATM Cell Header, while the CPCS-UU is transparently transported through the ATM network.
  - AAL-UNITDATA.indication primitive used by the CPCS to inform the SSCS of the arrival of a CPCS-PDU. The parameters include those for the send data request primitive, in addition to a Reception Status parameter to inform the SSCS whether the data being delivered is corrupted.
- For Streaming Mode service:
  - In streaming mode, the AAL-UNITDATA.request primitive data structure contains the parameters from the equivalent Message Mode primitive along with a More parameter to indicate whether there are additional IDUs required to form a complete CPCS-SDU. At this writing, the behavior of the streaming mode service with respect to the CPCS-LP, and CPCS-CI parameters is not specified in recommendation I.363.
  - AAL-UNITDATA.indication includes the Reception Status parameter.

The AAL 5 SAP address has a one-to-one association with an ATM Layer SAP address.

5.5 Summary

Table 3-6 presents a mapping of the AAL types to service classes. AAL 1 is clearly the only AAL that is able to provide circuit emulation due to the ability of this adaptation layer to carry end-to-end timing information. However, AAL 1 is not well suited for carrying VBR traffic. A potential issue in the use of AAL 1 to carry low bit rate sources, such as voice traffic, is that of delay; it takes approximately 6ms to accumulate enough voice samples to fill an ATM cell. An alternative, partial filling of ATM cells with only a few voice samples, reduces delay but has the disadvantage of wasting bandwidth.

AAL 2 allows packets from multiple higher layer connections to be multiplexed into a single ATM connection. Thus, voice samples from several connections could be multiplexed into a single ATM cell and demultiplexed at the termination of the VCC. By carrying short (less than an ATM Cell Payload) packets, AAL 2 allows for efficient use of bandwidth, and lower end-to-end delay than would be the case with AAL 1 using fully loaded ATM Cell Payloads. In addition, AAL 2 efficiently supports VBR sources. However, since AAL 2 does not carry timing information (à la AAL 1) it does not support Service Class B: VBR sources with end-to-end timing requirements.

While AAL 2 allows for more efficient use of the ATM cell payload than other AAL types for partial cell payloads, the advantages of AAL 2 are less apparent when carrying large PDUs. If AAL 2 is to be used to carry large PDUs, such would effectively obviate concerns about inefficient bandwidth use due to partial fill of the ATM cell payload. While

TABLE 3-6.

AAL to Service Class Mapping

	AAL 1	AAL 2	AAL 3/4	AAL 5
Service Class A	X			
Service Class B				
Service Class C		X	X	X
Service Class D			X	

the ability of AAL 2 to multiplex traffic from multiple VCCs into a single ATM Layer connection is a capability that it shares with AAL 3/4, the header overhead in an ATM Cell Payload associated with AAL 2 is, at best, no less than that of AAL 3/4 and is clearly greater than that of AAL 5.

While AAL 2 is capable of providing both VBR audio/video and VBR data services, AAL 5 is better suited to both for large packet size [here we will define "large" to mean greater than 12 ATM cell payloads, since at this point the number of AAL 2 SAR-PDU overhead octets is guaranteed to be greater than the size of the PAD field in the AAL 5 CPCS-PDU].

AAL 1, AAL 2, and AAL 5 are targeted for connection-oriented services. AAL 3/4 has provisions to support connectionless services. AAL 3/4 may also be used in connection-oriented VBR audio/video and VBR data services, but because of AAL 3/4 SAR-PDU overhead, AAL 5 provides a better alternative. The multiplexing feature of AAL 3/4 may be used in multiprotocol over ATM applications; however, the same capability is possible using AAL 5 if a protocol identifier is encapsulated within the AAL-SDU for processing by a higher layer protocol [15].

Thus, the primary application for AAL 3/4 is in connectionless data services. That said, there is nothing to prevent AAL 5 from being used to implement connectionless data services. In fact, as we will see in Chapter 8, IP over ATM Networks, there are a number of approaches defined to carry IP (a connectionless protocol) over ATM using AAL 5.

In conclusion, while there are four defined ATM Adaptation Layers, most ATM applications could be served by just two: AAL 1 and AAL 5.

References

1. Stallings, William, *Data and Computer Communications,* 3d Ed., Macmillan Publishing Co., 1991.

2. B-ISDN Protocol Reference Model and its Application, International Telecommunication Union, Recommendation I.321, April 1991.

3. ATM User-Network Interface Specification, Version 3.1, ATM Forum, 1994.

4. B-ISDN User-Network Interface—Physical Layer Specification, International Telecommunication Union, Recommendation I.432, March 1993.

5. Synchronous Multiplexing Structure, International Telecommunication Union, Recommendation G.709, March 1993.

6. ATM Cell Mapping Into Plesiochronous Digital Hierarchy (PDH), International Telecommunication Union, Recommendation G.804, November 1993.

7. B-ISDN ATM Layer Specification, International Telecommunication Union, Recommendation I.361, March 1993.

8. B-ISDN ATM Adaptation Layer Functional Description, International Telecommunication Union, Recommendation I.362, March 1993.

9. B-ISDN ATM Adaptation Layer Specification, International Telecommunication Union, Recommendation I.363, March 1993.

10. B-ISDN ATM Adaptation Layer Type 2 Specification, International Telecommunication Union, Recommendation I.363.2, September 1997.

11. Segmentation and Reassembly Service Specific Convergence Sublayer for the AAL Type 2, International Telecommunication Union, Recommendation I.366.1, September 1997.

12. B-ISDN ATM Adaptation Layer—Service Specific Connection Oriented Protocol (SSCOP), International Telecommunication Union, Recommendation Q.2110, August 1994.

13. Support of Broadband Connectionless Data Service on B-ISDN, International Telecommunication Union, Recommendation I.364, March 1993.

14. Comer, Douglas, *Internetworking with TCP/IP Vol. 1, 2d Ed.* Prentice-Hall, Inc., 1991.

15. Heinanen, J., *Multiprotocol Encapsulation over ATM Adaptation Layer 5,* Internet Engineering Task Force, RFC 1483, July 1993.

CHAPTER 4

ATM Signalling

1. Introduction

In this chapter we continue our discussion on ATM protocols to discuss the Control plane of the ATM Protocol Reference model introduced in the last chapter. The control plane is used for ATM signalling. In this chapter we may build upon the preceding chapter because the Adaptation Layer portion of ATM signalling is an AAL five-service specific convergence sublayer.

Figure 4-1 shows the interfaces across which signalling information passes:

- Private UNI—the interface between ATM terminal equipment and an ATM switch within a private ATM network.
- Private NNI—the interface between ATM switches within the private ATM network.
- Public UNI—the interface between the customer premises and the public carrier ATM network.
- NNI—the interface between ATM switches within a public carrier ATM network.
- B-ICI—the interface between ATM switches between the ATM networks of two different public carriers.

2. Signalling Protocol Architecture

The Signalling Protocol Architecture for ATM is defined by a Signalling AAL (SAAL) and one or more interface-specific Higher Layer protocols [1][2]. The SAAL consists of a Service Specific Convergence Sublayer (SSCS) on top of the AAL 5 CPCS and SAR. As shown in Figure 4-2, the SAAL consists of the following components:

- Common Part Convergence Sublayer: AAL 5.
- Segmentation and Reassembly Sublayer: also AAL 5.
- Service Specific Connection Oriented Protocol: SSCOP provides assured data transfer of signalling information between signalling endpoints.

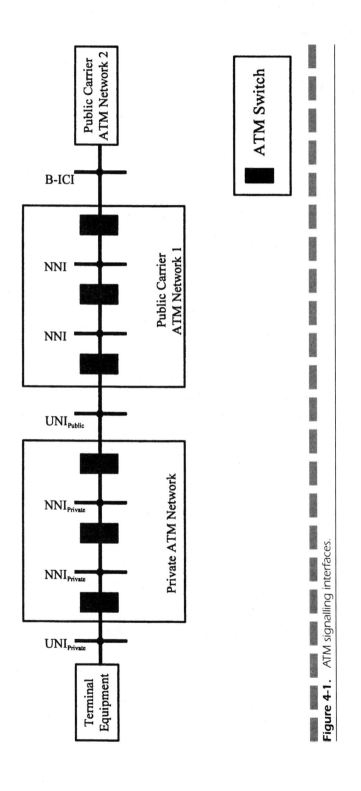

Figure 4-1. ATM signalling interfaces.

Figure 4-2.
Signalling AAL
(UNI and NNI).

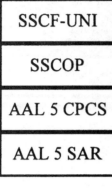
(a) UNI SAAL

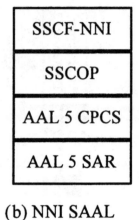
(b) NNI SAAL

- Service Specific Coordination Function: The SSCF maps SSCOP services to the needs of the Higher Layer protocol. There are two different SSCFs:
  - Service Specific Coordination Function at the UNI: SSCF-UNI maps SSCOP services to the needs of the Higher Layer protocols for ATM access signalling at the UNI (Figure 4-2a).
  - Service Specific Coordination Function at the NNI: SSCF-NNI maps SSCOP services to the needs of the Higher Layer protocols for signalling between ATM network nodes (Figure 4-2b).

In this section we present the signalling protocol architecture used at each of the interfaces shown in Figure 4-1. Since the signalling protocol architecture is the same between the Physical Layer and the AAL 5 CPCS, the discussion will focus on higher protocol entities.

In discussing higher layer protocol entities we should note that the term "Layer 3" has a different connotation when used to describe telephony systems vis-a-vis the OSI Protocol Reference Model. The OSI Layer 3 refers to the Network Layer entity that is responsible for routing and relaying of messages received from higher layer entities. In telephony, however, "Layer 3" in the control plane contains aspects of OSI Layers 3 through 7 [14]. Thus, when we refer to "Higher Layer," or "Layer 3" in this chapter, we are referring to an entity which provides the functions of Layers 3 through 7 of the OSI PRM.

2.1 Overview of Signalling Architectures

2.1.1 Private UNI

Figure 4-3a shows the signalling protocol architecture at the Private UNI:

- SSCOP
- SSCF-UNI
- Q2931: this is the B-ISDN network layer access protocol for access signalling at the UNI.

2.1.2 Private NNI

Figure 4-3b shows the signalling protocol architecture at the Private NNI:

- SSCOP
- SSCF-UNI
- Private Network to Network Interface: PNNI is a protocol for signalling and routing between network nodes in a private ATM network. PNNI is described in Chapter 8, ATM Routing.

2.1.3 Public UNI

The Public UNI signalling architecture is the same as that described for the Private UNI. This allows ATM terminal equipment to attach to either a Private or Public ATM network without modification.

2.1.4 NNI

Figure 4-3c shows the signalling protocol architecture at the NNI.

- SSCOP
- SSCF-NNI

Figure 4-3.
Signalling protocol architecture.

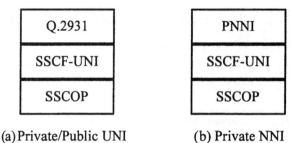

(a) Private/Public UNI   (b) Private NNI

(c) Public NNI   (d) B-ICI

- Broadband ISDN User Part: B-ISUP provides signalling capabilities which support ATM bearer services on Signalling System Number 7 (SS7) networks.
- Broadband Message Transfer Part at Level 3: BMTP-3 is a protocol for the transfer of messages between signalling endpoints in an SS7 network.

2.1.5 B-ICI

Figure 4-3d shows the signalling protocol architecture at the B-ICI:

- SSCOP
- SSCF-NNI
- BMTP-3 in Associated Mode: this is a simplified version of MTP-3.
- B-ISUP

2.2 ATM Connection Establishment

In this chapter we will describe the procedures for the creation and management of two types of ATM connections:

- Signalling VCC
- User Traffic SVC

2.2.1 Signalling VCC

Before on-demand SVCs may be established to carry cell traffic in the User Plane, a Signalling VCC must be established to carry signalling messages in the Control Plane. This requires establishment of signalling VCCs, which are to carry control plane messages that allow for the establishment and release of SVCs. The signalling VCC establishment process is performed by the SAAL, which establishes a signalling connection that supports assured data transfer between signalling endpoints. When interfacing ATM to SS7 networks, BMTP-3 is used in conjunction with SAAL.

There are two types of signalling associated with the signalling VCC:

- Associated Signalling: In associated signalling, the signalling VCC controls SVC connection establishment within the same virtual path (as identified by a VPI value) that carries the signalling VCC. This is not to be confused with Associated Mode Signalling.
- Nonassociated Signalling: In this case, the signalling VCC may control SVC connection establishment in other virtual paths carried over the same physical interface.

The selection of associated or nonassociated signalling is made at the time connection establishment is requested. Associated signalling is not supported in ATM Forum UNI 3.1 or UNI 4.0. Therefore, signalling messages must explicitly specify a value for VPCI.

2.2.2 User Traffic SVC

Once a signalling VCC is established, higher layer protocols may use this connection to establish and release connections that carry user traffic.

2.3 Signalling ATM Adaptation Layer

2.3.1 Service Specific Connection Oriented Protocol

The purpose of SSCOP in the SAAL is to provide a mechanism for the assured transfer of information between signalling endpoints [3]. The functions performed by the SSCOP include the following:

- Connection Control—performs connection establishment and release between peer (sender and receiver) SSCOP entities.
- Transfer of Data—allows the transfer of data between peer SSCOP entities. SSCOP supports both assured and nonassured information transfers.
- Sequence Integrity—SSCOP SDUs are received in the same order that they were transmitted from the sender.
- Selective Retransmission—the SSCOP at the receiving end can detect missing SSCOP SDUs and can request that the sender retransmit only those missing SDUs.
- Flow Control—the receiver is able to control the rate at which SSCOP SDUs are transmitted from the sender
- Local Data Retrieval—allows the local SSCF to retrieve SDUs (in sequence) which have not been transmitted.

2.3.1.1 Connection Control As shown in Figure 4-4, the SSCOP connection establishment process consists of an AA-ESTABLISH request from the SSCF entity in the originating signalling endpoint which produces an indication when the message is received at the SSCF entity in the terminating endpoint. In reply the SSCF in the terminating endpoint sends a response to the originator indicating whether the connection request is accepted or refused. When received by the originator, this response produces a confirmation that is sent to the SSCF entity. The originator inspects the contents of the return message to determine whether connection establishment has been successful. If a connection has been successfully established, the signalling endpoints may exchange information using assured data transfer procedures.

Chapter 4: ATM Signalling

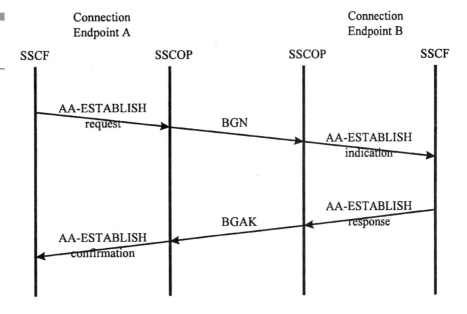

Figure 4-4. SSCOP connection establishment.

The release of the SSCOP connection involves an AA-RELEASE request which may originate from either signalling endpoint that produces an indication at the peer signalling endpoint that is sent to the SSCF entity (Figure 4-5). The release request message may be produced by the SSCF, or the SSCOP may initiate connection release (e.g., in case of an unrecoverable protocol error). When the connection release request is received by the peer, the SSCOP entity in the peer sends a return message. When this return message is received by the originator, a confirmation is generated which is sent to the SSCF entity. This confirmation is sent to the originating side SSCF entity regardless of whether the release request originated within the SSCF or the SSCOP.

The SSCOP interfaces between the AAL 5 CPCS and the SSCF (the SSCOP user). For connection control, the interface between the SSCOP and SSCF includes of the following primitives:

- AA-ESTABLISH: which is used to establish a point-to-point connection for assured information transfer between signalling endpoints. This message includes an SSCOP user-to-user (SSCOP-UU) PDU which coveys information between SSCF entities at the signalling endpoints. SSCOP-UU PDUs in an AA-ESTABLISH may be one of the following types:

Figure 4-5.
SSCOP connection release.

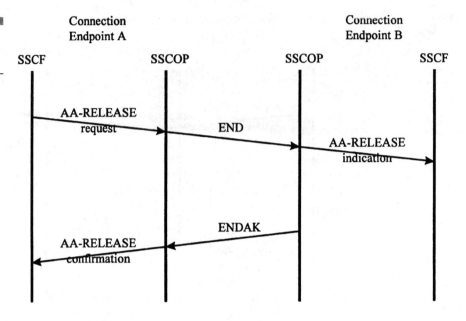

- BGN: indicating that the AA-ESTABLISH message is a request to establish an SSCOP connection.
- BGAK: which indicates that the connection establishment request has been accepted by the peer signalling entity.
- BGREJ: indicates that the connection establishment request has been rejected by the peer signalling entity.

- AA-RELEASE: terminates a point-to-point connection for assured information transfer. This message includes an SSCOP-UU PDU and an indication of the source of the release message (SSCF or SSCOP). If the source is SSCOP, the SSCOP-UU PDU is ignored. SSCOP-UU PDUs in an AA-RELEASE may be one of the following types:
  - END: indicates that the SSCOP connection is to be released.
  - ENDAK: confirms connection release.

2.3.1.2 Transfer of Data, Sequence Integrity, Selective Retransmission, Flow Control, and Error Recovery SSCOP supports both assured and nonassured data transfer. In either case the data transfer process consists of a request from the SSCF entity in the transmitting side which produces an indication at the receiving side SSCF entity.

Assured Data Transfer The connection establishment procedure of the last section sets up a number of state variables in the SSCOP entities in the signalling endpoints to support assured data transfer. The assured data transfer method was described in Chapter 3 in the section describing the Service Specific Assured Data Transfer Sublayer of AAL Type 2.

The interface between the SSCOP and the SSCF for assured data transfer includes an AA-DATA primitive which indicates that the assured data transfer method is being invoked. This message includes a Message Unit (MU) SDU, which may be one of the following types:

- Sequenced Data (SD): is used to transfer a sequentially numbered PDU containing information from the SSCF entity.
- Poll: used to request status information from the peer SSCOP entity.
- Solicited Status (STAT): is used to respond to a Poll PDU.
- Unsolicited Status (USTAT): is sent to the peer SSCOP entity when one or more missing SSCOP PDUs is detected.

There are a number of error conditions that may occur in assured data transfer mode, e.g., such as may result when state variables between the peer SSCOP entities are not in agreement. SSCOP also supports procedures for recovery from protocol errors related to sequence number problems. The following primitives between the SSCOP and SSCF support resynchronization and protocol error recovery:

- AA-RESYNC: used to resynchronize the SSCOP connection.
- AA-RECOVER: used during recovery from protocol errors.

Nonassured Data Transfer Nonassured data transfer allows connectionless data transfer between peer SSCOP entities. No state variables are created or modified as a result of nonassured data transfer. The nonassured data transfer method does not employ selective retransmission, sequence integrity, or flow control mechanisms.

The interface between the SSCOP and the SSCF for nonassured data transfer includes an AA-UNITDATA primitive, which indicates that data transfer is to be effected using the nonassured data transfer method. This message includes an MU, which is an Unnumbered Data (UD) SDU that contains information from the SSCF entity.

2.3.1.3 Local Data Retrieval In some instances a transmitting SSCF entity will request retrieval of SSCOP SDUs from the local SSCOP that have not yet been sent to the peer signalling endpoint. This occurs in the SSCF-NNI only and is used in support of BMTP-3 procedures, which perform message retrieval as part of a recovery procedure when rerouting signalling message traffic. Only PDUs sent using the assured data transfer method are retrieved.

In the local data retrieval procedure, the SSCF entity sends a request message to the local SSCOP entity, instructing that the SSCOP return one or more untransmitted SSCOP SDUs to the SSCF. The corresponding indication from the SSCOP to the SSCF contains the appropriate untransmitted SDUs.

To support local data retrieval, the interface between the SSCOP and SSCF includes the following primitives:

- AA-RETRIEVE: a request message contains a retrieval number (RN) parameter that indicates to the local SSCOP entity which SDUs are to be retrieved. An indication message contains an assured data transfer MU.
- AA-RETRIEVE-COMPLETE: an indication sent from the SSCOP to the SSCF to signal that local data retrieval has been completed.

2.3.1.4 Interface between SSCOP and CPCS The interface between SSCOP and the AAL 5 CPCS is as described in Chapter 3 in the description of the AAL 5 SAP. SSCOP uses the Message Mode service from the AAL 5 CPCS.

2.3.2 Service Specific Coordination Function at the UNI

The purpose of the SSCF-UNI is to map the services of the SSCOP to the Q.2931 protocol for access signalling across the UNI [4]. The SSCF-UNI provides an SAP to the Layer 3 protocol as defined by a set of primitives. There is a one-to-one correspondence between a connection at the AAL-SAP, and one at the ATM-SAP.

The following services are provided at the SSCF-UNI SAP:

- Assured Data Transfer: for SDUs over point-to-point ATM connections.

Chapter 4: ATM Signalling

- Connection Establishment and Release for Assured Data Transfer.
- Nonassured Data Transfer: for unacknowledged transfer of SDUs.

As shown in Figure 4-6, the connection establishment process in SSCF-UNI is initiated by an AAL-ESTABLISH request from the Q2931 entity in the originating signalling endpoint which produces an indication when the message is received at the Q2931 entity in the terminating endpoint. The receiving SSCF entity sends a return message that generates a confirmation at the originating Q2931 entity.

Once the connection is established, assured data transfer may occur between peer Q2931 entities.

The SSCF-UNI maps SAP primitives to primitives at the SSCF-UNI/SSCOP interface:

- AAL-ESTABLISH maps to the AA-ESTABLISH primitive in SSCOP.
- AAL-RELEASE: maps to AA-RELEASE.
- AAL-DATA: maps to AA-DATA.
- AAL-UNIT_DATA: maps to AA-UNITDATA.

2.3.3 Service Specific Coordination Function at the NNI

The SSCF-NNI maps SSCOP services to network node signalling over SS7 networks as defined in Q704 [5][6][7]. The services required to support

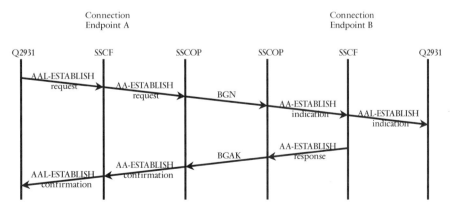

Figure 4-6.
SSCF-UNI connection establishment.

Q704 require SSCF-NNI to support a wider range of services than is the case in SSCF-UNI.

The following services are provided at the SSCF-NNI SAP:

- Assured Data Transfer: for SDUs over point-to-point ATM connections.
- Connection Establishment and Release for Assured Data Transfer.
- Local SDU Retrieval: allows the local Q704 entity to retrieve SDUs that have been submitted to the SSCF but not yet transmitted to the peer Q704 entity.
- Signalling Link Error Monitoring: SSCF-NNI supports Q704 procedures to remove and restore signalling links to an in-service state.
- Flow Control: SSCF-NNI provides mechanisms that allow for detection of local congestion on the signalling link and control of SDU traffic flow.

2.3.3.1 CONNECTION ESTABLISHMENT AND RELEASE

Connection Establishment At the NNI, the connection establishment process involves changes in link status. Since a link is a viewable management object this process involves interaction with the Layer Management entity (in the M-plane). Assuming that the signalling link between signalling endpoints is in the out-of-service state, the connection establishment process, shown in Figure 4-7, begins with the transmitting side B-MTP 3 entity issuing an AAL-START request across the SSCF-NNI SAP. The SSCF-NNI, in turn, issues an AA-ESTABLISH request to the SSCOP. The SSCF-NNI also sends a MAAL-REPORT indication to the Layer Management entity indicating that the signalling link is being restored to service (a process referred to as the *alignment stage*).

Upon receipt of an AA-ESTABLISH indication at the receiving SSCF-NNI entity, the local Layer management entity is sent an MAAL-PROVING indication to inform the management entity that the SSCF entity is to begin link proving. The receiving SSCF-NNI entity also sends an AA-ESTABLISH response to acknowledge connection establishment. The originating SSCF-NNI entity receives an AA-ESTABLISH confirmation and sends a MAAL-PROVING indication to its local Layer management entity. At this point, both connection endpoints are performing link proving.

Link proving is a process by which SSCF peer entities send proving PDUs. The purpose of this procedure is to monitor the quality of the

Chapter 4: ATM Signalling

link according to its ability to provide error-free delivery of PDUs between connection endpoints. Proving PDUs are a special type of SSCF-NNI PDU that is not transparent to the SSCF. During the link proving procedure each SSCF-NNI entity sends proving PDUs in AA-DATA request messages to the peer entity. Timers at each SSCF-NNI entity are used to detect lost-proving PDUs.

If link performance is acceptable after the requisite number of proving PDUs are sent, the proving process ends, and an SSCF-NNI PDU is sent to indicate that the link is in-service for signalling traffic. The local Layer management entity is also sent an MAAL-STOP_PROVING indication that link proving has ended.

When the receiving SSCF-NNI entity receives the SSCF-NNI PDU in an AA-DATA indication, an AAL-IN_SERVICE indication is sent to

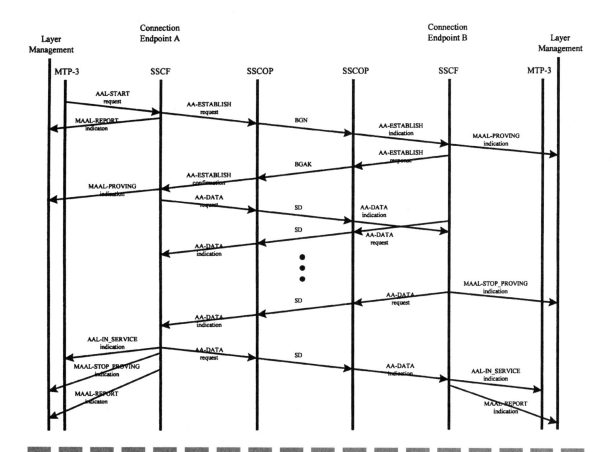

Figure 4-7. SSCF-NNI connection establishment.

the Q.704 entity to indicate that the link is in service and ready to transfer signalling messages between peer Q.704 entities. In addition, a MAAL-REPORT indication is sent to the Layer management entity, which also indicates that the link is in the in-service state.

Connection Release The connection release process is shown in Figure 4-8. It begins when a BMTP-3 entity issues an AAL-STOP request. The local SSCF-NNI entity generates an AA-RELEASE request containing an SSCF-NNI PDU indicating that the link is to be taken out of service. The local Layer management entity is also sent an MAAL-REPORT indicating local release of the signalling link and that the link is to be placed in out-of-service state. The BMTP-3 entity has the opportunity to retrieve signalling messages which have been submitted to the local

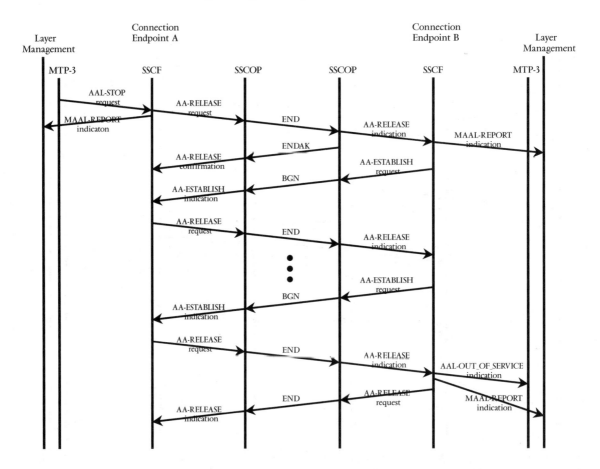

Figure 4-8. SSCF-NNI connection release.

SSCF-NNI entity, but not yet transmitted to the peer entity over the out-of-service link (Local SDU Retrieval).

The receiving SSCOP entity receives an END PDU and generates an ENDAK acknowledgment to the originating peer SSCOP entity. This results in an AA-RELEASE confirmation being sent to the originating SSCF-NNI entity. The receiving SSCOP entity also sends an AA-RELEASE indication to the local SSCF-NNI entity. Based upon the information in the received SSCF-NNI PDU, the receiving SSCF-NNI entity sends an MAAL-REPORT indication to the local Layer management entity indicating that the peer SSCF-NNI entity has released the connection.

The receiving side SSCF-NNI will then attempt an alignment procedure to reestablish the connection by sending an AA-ESTABLISH request to the peer entity. Since the peer is now out of service, it will return an AA-RELEASE request in response. The SSCF-NNI entity receiving the original AA-RELEASE indication will continue to reattempt connection reestablishment until a local timer expires. At that point, the SSCF-NNI will inform the local BMTP-3 entity that the signalling link is out of service by sending an AAL-OUT_OF_SERVICE indication. In addition, an MAAL-REPORT indication is sent to the Layer management entity to report that the local link is being taken out of service due to failure of the alignment procedure. The SSCF-NNI also sends an AA-RELEASE request to the peer SSCF-NNI entity containing an SSCF-NNI PDU, which indicates that the attempted alignment procedure was not successful.

For connection establishment, the SSCF-NNI SAP provides the following primitives to the BMTP-3:

- AAL-START: start process of connection establishment.
- AAL-STOP: inhibit SSCF-NNI peer to SSCF-NNI peer communication.
- AAL-IN-SERVICE: signalling link is available for transport of signalling messages.
- AAL-OUT_OF_SERVICE: signalling link not available.
- AAL-EMERGENCY: request reduction in rate of link proving PDUs.
- AAL-EMERGENCY_CEASES: return to normal link proving.

2.3.3.2 Assured Data Transfer SSCF-NNI uses the assured data transfer method exclusively. The SAP provides the following primitives to the BMTP-3:

- AAL-MESSAGE_FOR_TRANSMISSION maps to the AA-DATA request primitive in SSCOP.
- AAL-RECEIVED_MESSAGE maps to the AA-DATA indication SSCOP primitive.

2.3.3.3 Local SDU Retrieval

When signalling traffic between connection endpoints in an SS7 network is rerouted over a link other than the current one (referred to as a *changeover procedure*), the SSCF-NNI supports procedures which allow the BMTP-3 entity to retrieve SDUs which have not been sent and acknowledged by the peer signalling entity. This reduces the possibility of a message being lost or received multiple times. Local SDU retrieval is a two-step process:

1. Ensure that all SDUs sent from the peer BMTP-3 entity have been received.
2. Retrieve any SDUs submitted from the local BMTP-3 entity that have not been received by the peer.

SSCF-NNI SAP provides the following primitives to the BMTP-3 for local data retrieval:

- AAL-RETRIEVE_BSNT: requests/indicates that BSNT (Backward Sequence Number Threshold) value be retrieved. The BSNT should be equal to the sequence number of the last PDU received from the peer SSCF-NNI entity.
- AAL-BSNT: delivers the BSNT value to the local BMTP-3 entity.
- AAL-BSNT_NOT_RETRIEVABLE: notifies BMTP-3 that BSNT value could not be retrieved.
- AAL-RETRIEVE_REQUEST_AND_FSNC: requests retrieval of SDUs that have not been acknowledged by peer.
- AAL-RETRIEVAL_COMPLETE: retrieval of unacknowledged SDUs is complete.

2.3.3.4 Flow Control

Q.704 maintains procedures for detection of traffic congestion on signalling links. The SSCF-NNI provides primitives at the SAP that allow the BMTP-3 entity to communicate link congestion status:

- AAL-LINK_CONGESTED: indicates local congestion on signalling link.
- AAL-LINK_CONGESTION_CEASED: indicates local congestion has ceased.

2.4 Public UNI Signalling

Public UNI signalling consists of a set of procedures for call control and bearer control between a subscriber and a public B-ISDN network at the T_B interface. The capabilities currently supported by UNI signalling include:

- Basic signalling functions
    - Connection establishment: allows a calling party to request connection to a called party and to provide information pertaining to the connection and its characteristics. The called party may respond to the connection establishment request and may also provide connection-related information.
    - Connection clearing: allows a party involved in an established connection to end participation in the connection.
    - Out-of-band signalling: call control information uses a different VCC than is used for exchanging information between end parties. The VCI used within any VPI for signalling is a *well-known* value.
- On-demand connection characteristics
    - Point-to-point connection: connection involving two communicating endpoints.
    - Point-to-multipoint: connection involving more than two communicating endpoints. One of the parties serves as the root node while the other parties are leaf nodes.
    - Unidirectional/bidirectional: information transfer on an established connection may be one-way or duplex.
    - Symmetrical/asymmetrical bandwidth: a bidirectional connection may support bit rates which are equal in each direction (symmetrical), or unequal (asymmetrical).
- Specification of QoS Class
- AAL parameter negotiation: a calling party may request a set of connection parameters and may receive an indication of an alternate set of parameters proposed by the called party. There are rules that dictate allowable responses to a request.
- VPCI/VPI/VCI assignment

UNI signalling procedures are described in ITU Recommendations Q.2931 and Q.2971 for point-to-point and point-to-multipoint call connections respectively [8][9]. The corresponding ATM Forum standards are defined in ATM UNI specification version 3.1, and ATM UNI version 4.0 [10][11]. While the respective standards documents from the ITU and ATM Forum are substantially in agreement, there are points of departure as will be discussed in the following sections.

Prior to any transfer of signalling messages across the UNI, a connection must be established between signalling endpoints, which supports assured data transfer. Signalling messages are carried within the AAL-DATA primitive defined at the SSCF-UNI SAP.

2.4.1 Point-to-Point Connection

Connection Establishment Figure 4-9 shows the call control procedures for point-to-point connection establishment. The calling party initiates call establishment by transferring a SETUP message to the network across the Public UNI using the Signalling VCC. In B-ISDN signalling the SETUP message contains all information needed by the network to establish the connection including:

- Call reference: a unique number used by the network to identify the call.
- Called party address
- QoS class
- ATM Traffic Descriptors (which indicate whether this is a unidirectional/bidirectional, asymmetrical/symmetrical connection)
- AAL Parameters

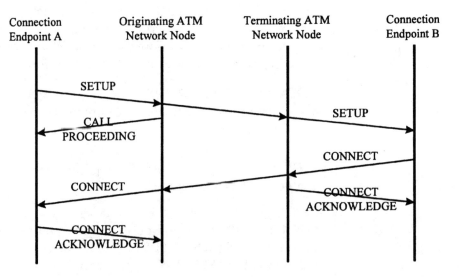

Figure 4-9. Point-to-point connection establishment.

Chapter 4: ATM Signalling

A point of departure between the ITU and ATM Forum in connection establishment procedures is that Q2931 allows the connection endpoint to specify a connection identifier (VPCI and VCI values) in the SETUP message. This capability, which would support direct signalling between connection endpoints, is not supported in ATM Forum UNI 3.1. According to UNI 3.1, the connection endpoint must signal to the network across the UNI, which assigns the connection identifier. In ATM Forum UNI 4.0, however, direct endpoint-to-endpoint signalling is supported.

If the connection establishment request is accepted by the network, a CALL PROCEEDING message is sent back to the calling party to acknowledge receipt of the SETUP message and to indicate that the call is being processed.

At the UNI of the called party, the network indicates arrival of the connection establishment request by transferring a SETUP message to the called party across the UNI. If the SETUP message is accepted by the called party, a CONNECT message is sent to the network. This CONNECT message may contain connection parameters (e.g., AAL parameters) which are different from those of the SETUP message. These constitute a negotiation from the called party to the calling party for a modified set of connection parameters. In response to the CONNECT message, the network sends a CONNECT ACKNOWLEDGE message.

At the calling party UNI, the network transfers the CONNECT message to the calling party which is the indication to the calling party that the SVC has been established. The calling party may inspect the CONNECT message for any modifications in parameters from the original SETUP message. If there are, and the alternate parameter values are acceptable, the calling party uses these values for the duration of the connection. Otherwise, the calling party initiates connection-clearing procedures to end the connection.

Connection Clearing Connection clearing may be initiated by either party or the network and is shown in Figure 4-10. If initiated by one of the parties, the connection-clearing process begins with a RELEASE message being sent to the network using the signalling VCC. The network responds with a RELEASE COMPLETE. This is the indication that participation of the requesting party in the SVC has been ended. It does not mean, however, that the remote party has been disconnected from the SVC.

Across the remote UNI, the network sends a RELEASE message to the remote party. In response, the remote party sends a RELEASE COMPLETE, which indicates to the network that the remote party is no longer a participant in the SVC. At this point, the SVC is terminated.

Figure 4-10.
Point-to-point connection release.

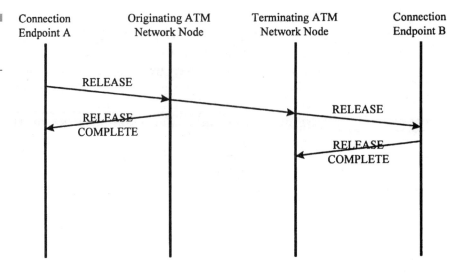

2.4.2 Point-to-Multipoint Connection

2.4.2.1 ROOT INITIATED JOIN

Addition of Parties Point-to-multipoint connection is described in ITU recommendation Q2971 and UNI 3.1/UNI 4.0. The first leg of a point to multipoint connection uses the procedures of point-to-point connection as described above, but the setup message contains information that indicates that a point-to-multipoint connection is being established. The initiator of the connection in this case is known as the *root*. Every other participant in the connection is referred to as a *leaf*. In point-to-multipoint connection, a leaf is not able to negotiate connection parameters with the root. As the root adds leaves to the connection, each request references the call reference that was assigned at the time that the point-to-multipoint connection was established.

At the time of this writing Q2971 procedures require that the root initiate addition of any leaves to the connection. ATM UNI 4.0, however, allows a leaf to initiate joining an existing multipoint connection (referred to as a *leaf-initiated join*, or LIJ).

In Q2971 procedures, shown in Figure 4-11, the root sends an ADD PARTY message to the network indicating a called party to be added to an existing connection. In response, the network sends a SETUP message to the called party. As in point-to-point connection, the called party indicates acceptance of the connection by sending a CONNECT message to the network, and receives a CONNECT ACKNOWLEDGE message from the network in response. The root receives an ADD PARTY ACKNOWLEDGE message indicating the successful addition of the leaf to the existing multipoint connection.

Dropping Parties Either the root or a leaf may initiate dropping of that leaf from a multipoint connection. The case of a root-initiated drop is shown in Figure 4-12. The root sends a DROP PARTY message to the network, receiving a DROP PARTY ACKNOWLEDGE from the network. The leaf is released using Q2931 procedures.

A leaf-initiated drop uses the same procedures as Q2931 involving an exchange of RELEASE and RELEASE COMPLETE messages. Across the root UNI, the network sends a DROP PARTY message. The root responds with a DROP PARTY ACKNOWLEDGE.

The root may also drop from the connection; however, if the root drops from the connection, all leaves are also dropped. In this case, the root uses Q2931 connection clearing procedures (RELEASE/RELEASE COMPLETE). In response, the network will initiate Q2931 connection-clearing procedures at each leaf of the multipoint connection.

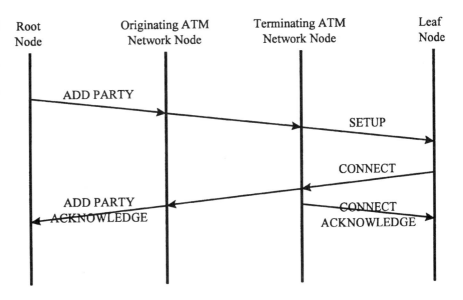

Figure 4-11. Point-to-multipoint connection add party (root initiated).

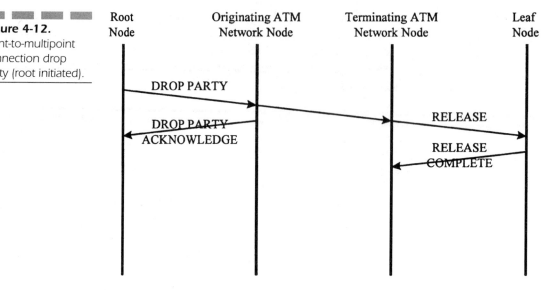

Figure 4-12.
Point-to-multipoint connection drop party (root initiated).

2.4.2.2 Leaf-Initiated Join ATM UNI 4.0 provides the capability for a leaf-initiated join. This feature obviates the root having to perform all actions regarding the addition of parties to a point-to-multipoint connection. There are two varieties of LIJ:

- Leaf-Prompted Join without Root Notification: this allows a leaf node to join and drop from a point-to-multipoint connection with no notification to the root that a party has joined or dropped from the connection. Also referred to as Network-LIJ.

- Root-Prompted Join: this capability allows the leaf to issue a request to the root to join a point-to-multipoint connection. In response, the root may perform a root-initiated join as described in the previous section. Also referred to as Root-LIJ.

The selection of Network-LIJ, or Root-LIJ is determined by the root when it establishes the point-to-multipoint connection. The root identifies the point-to-multipoint connection with an LIJ Call Identifier at the time of call establishment. Any join or join request initiated by the leaf must reference this identifier in the connection request from the leaf.

Leaf-Prompted Join without Root Notification Figure 4-13 shows the Network-LIJ process. The leaf sends a LEAF SETUP REQUEST message to the network to indicate a request to join an existing multipoint connection. Even though the root node is not notified of this request, it is still indicated as the called party in the request. The network responds by sending a SETUP message to the calling leaf party. Connection establishment is completed using Q2931 connection procedures at the leaf UNI.

Root-Prompted Join Figure 4-14 shows the Root-LIJ process. In this case, the leaf sends a LEAF SETUP REQUEST to the network. The network forwards the message to the root. Addition of the leaf proceeds using Q2971 procedures.

Dropping Parties Dropping parties under leaf-initiated join follows Q2971 procedures.

2.5 Private Network to Network Interface Signalling

PNNI signalling is defined by the ATM Forum for signalling within private ATM networks [12]. The connection control procedures follow those described in Q2931. In addition to signalling, the PNNI specifica-

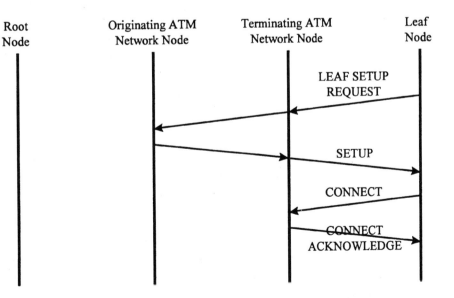

Figure 4-13. Point-to-multipoint connection add party (network-LIJ).

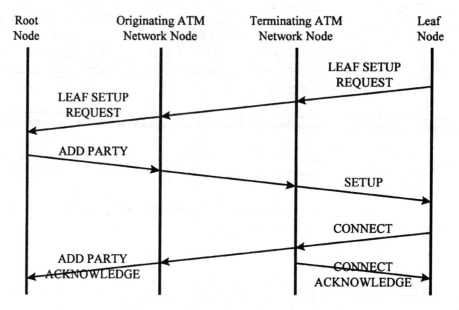

Figure 4-14.
Point-to-multipoint connection add party (root-LIJ).

tion also defines a routing protocol that allows an originating ATM node to determine a path through the network to a destination ATM node. PNNI routing is discussed in Chapter 8.

2.6 Broadband ISDN User Part

B-ISUP refers to the call procedures that are to be executed at the interfaces between public networks, or "exchanges" [13]. These procedures govern the transfer of information between exchanges involved in the completion of a connection between calling parties. B-ISUP does not specify procedures to be used within an exchange. These procedures include:

- Determination of whether there are resources available for the completion of the call.
- Selection of a route between the origination exchange and the destination exchange.
- Negotiation of connection identifiers (VPCI/VCI) between exchanges.
- Interworking with Narrowband ISDN endpoints.

2.6.1 B-ISUP Specification Model

The B-ISUP specification model used to describe B-ISUP procedures within a network node is based upon the OSI Application Layer Structure as described in ITU recommendation Q1400 [14]. Using this model, we may describe the generic B-ISUP software architecture of a network node as consisting of:

- Application Process (AP): in this discussion the application process would represent the signalling protocols supported at the node.
- Nodal Functions: the functions within a network node which support a specific protocol (e.g., B-ISUP).
- Application Entity (AE): a function used by the AP to communicate with a peer in another node. An AP may use multiple AEs of different types. Each instance of an AE is referred to as an Application Entity Instance (AEI) of a given type, e.g., an AEI which supports B-ISUP Nodal functions is a B-ISUP AEI.
- Application Service Entity (ASE): a subprocedure within an AE that performs a function or set of functions that, with other ASEs, define the set of capabilities of the AE. The ASEs specific to B-ISUP include:
  - Call Control ASE (CC): responsible for procedures that maintain the current state of the call (in progress, active, released, etc.).
  - Bearer Connection Control ASE (BCC): procedures which set up and tear down links between adjacent exchanges in the SVC call path.
  - Maintenance Control ASE (MC): procedures that handle management plane functions.
  - Unrecognized Information ASE (UI): procedures that handle unrecognized messages.
- Association Control Service Element (ACSE): a special ASE that sets up and releases associations between peer AEs. The connection established between signalling entities for assured data transport is an example of an association. Thus, the ACSE would be responsible for establishment and release of the signalling connection.
- Single Association Control Function (SACF): the set of rules which control interactions and joint use of ASEs.

- Single Association Object (SAO): the set of functions that communicate over a single association to a peer. In a B-ISUP AE the SAO is one of the following types:
  - Incoming call and connection control: consisting of BCC, CC, MC, and UI ASEs, and an SACF.
  - Outgoing call and connection control: consisting of BCC, CC, MC, UI, and SACF.
  - Maintenance: consisting of MC, UI, and SACF.

When the application process in an exchange determines that B-ISUP procedures are required for call/connection control, it creates an instance of B-ISUP. The B-ISUP nodal functions, in turn, create AEIs for each required signalling association.

Figure 4-15 presents the B-ISUP specification model for an intermediate exchange. Included in this model is a Network Interface (NI) function that provides an interface between the local BMTP-3 entity and the AEIs within the exchange node. This model shows two AEIs: one for the incoming VC from the preceding exchange (from the originating exchange), and one for the outgoing VC to the succeeding exchange (toward the destination exchange). Each AEI is identified by a unique Signalling Identifier (SID) which is created when the AEI is created (i.e., during SVC establishment). The SID is used to label signalling messages between peer entities relating to the AEI. This SID (and the AEI) is deleted when the SVC is released.

2.6.2 Call Control and Bearer Connection Control

In the model shown in Figure 4-15 we are able to see a conceptual separation of call control and bearer control into separate functions. For example, when a SETUP message is received for a point-to-point connection, it contains information pertinent to both call control and bearer connection control. When receiving a SETUP request or indication, the SACF creates requests for the CC and BCC ASEs respectively. This procedure is consistent with the B-ISDN Signalling Capability Set 1 as defined by the ITU, in which call control and bearer control capabilities are combined.

Chapter 4: ATM Signalling

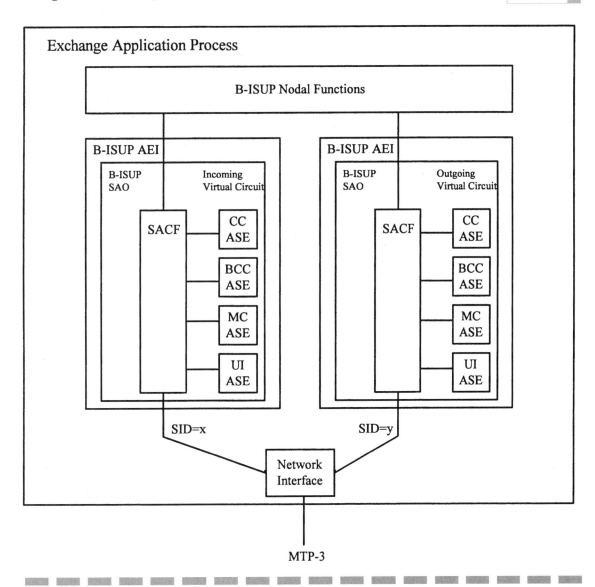

Figure 4-15. B-ISUP specification model.

The conceptual separation of call control and bearer control creates a framework for support of teleservices for which this separation is visible. In point-to-multipoint call connection, for example, we see a departure from combined call/bearer connection control paradigm in the leaf-initiated join case. Point-to-multipoint call connection separates bearer connection control from call control by allowing a leaf to join an existing connection.

2.7 Broadband Inter-Carrier Interface

The ATM Forum has defined a specification for the B-ICI [15]. This specification largely follows the BMTP-3 and B-ISUP procedures as defined by ITU recommendations Q704/Q2210, and Q2764 respectively. There are, however, points of departure between the ATM Forum and ITU. Differences to be noted include:

1. B-ICI makes no distinction between national and international exchanges.
2. B-ICI does not perform B-ISDN/N-ISDN interworking.
3. B-ICI supports BMTP-3 associated mode signalling only.

On the last point, BMTP-3 is defined to support complex SS7 networks in which signalling messages may be routed by one or more Signalling Transfer Points (STP) between network nodes [16][17]. This is referred to as quasi-associated mode signalling. B-ICI uses a simplified version of BMTP-3 in which B-ISUP signalling messages are transported directly between network nodes. This is referred to as *associated mode* signalling.

References

1. B-ISDN Signalling ATM Adaptation Layer (SAAL) Overview Description, International Telecommunication Union, Recommendation Q2100, August 1994.

2. B-ISDN Signalling ATM Adaptation Layer (SAAL) Layer Management for the SAAL at the Network Node Interface (NNI), International Telecommunication Union, Recommendation Q2144, October 1995.
3. B-ISDN ATM Adaption Layer—Service Specific Connection Oriented Protocol (SSCOP), International Telecommunication Union, Recommendation Q2110, August 1994.
4. B-ISDN Signalling ATM Adaption Layer—Service Specific Coordination Function for Support of Signalling at the User Network Interface (SSFC at UNI), International Telecommunication Union, Recommendation Q2130, August 1994.
5. B-ISDN ATM Adaption Layer—Service Specific Coordination Function for Support of Signalling at the Network Node Interface (SSFC at NNI), International Telecommunication Union, Recommendation Q2140, February 1995.
6. Signalling System No. 7—Signalling Network Functions and Messages, International Telecommunication Union, Recommendation Q704, March 1993.
7. Message Transfer Part Level 3 Functions and Messages Using the Services of ITU-T Recommendation Q2140, International Telecommunication Union, Recommendation Q2210, August 1996.
8. Broadband Integrated Services Digital Network (B-ISDN—Digital Subscriber Signalling System No. 2 (DSS 2)—User-Network Interface (UNI) Layer 3 Specification for Basic Call/Connection Control, International Telecommunication Union, Recommendation Q2931, February 1995.
9. Broadband Integrated Services Digital Network (B-ISDN) Digital Subscriber Signalling System No. 2 (DSS 2) User-Network Interface Layer 3 Specification for Point-to-Multipoint Call/Connection Control, International Telecommunication Union, Recommendation Q2971, September 1995 (Draft).
10. ATM User-Network Interface Specification, Version 3.1, ATM Forum, 1994.
11. ATM User-Network Interface (UNI) Signalling Specification, Version 4.0, ATM Forum, July 1996.
12. Private Network-Network Interface Specification Version 1.0 (PNNI 1.0), ATM Forum, March 1996.

13. Broadband Integrated Services Digital Network (B-ISDN)—Signalling System No. 7 B-ISDN User Part (B-ISUP)—Basic Call Procedures, International Telecommunication Union, Recommendation Q2764, February 1995.

14. Architecture Framework for the Development of Signalling and OA&M Protocols Using OSI Concepts, International Telecommunication Union, Recommendation Q1400, March 1993.

15. BISDN Inter Carrier Interface (B-ICI) Specification Version 2.0 (Integrated), ATM Forum, December 1995.

16. Introduction to CCITT Signalling System No. 7, International Telecommunication Union, Recommendation Q700, March 1993.

17. Stallings, William, *Data and Computer Communications*, 3d Ed. Macmillan Publishing Co., 1991.

18. Broadband Integrated Services Digital Network (B-ISDN)—Interworking Between Signalling System No. 7—Broadband ISDN User Part (B-ISUP) and Narrowband ISDN User Part (N-ISUP), International Telecommunication Union, Recommendation Q2660, February 1995.

CHAPTER 5

Operations Administration, Maintenance, and Provisioning

1. Introduction

In the preceding chapters we have described the layers of the B-ISDN Protocol Reference Model:

- Physical Layer
    - Transmission Path
    - Digital Section
    - Regenerator Section

- ATM Layer
    - Virtual Path
    - Virtual Channel

- ATM Adaptation Layer
    - Segmentation and Reassembly
    - Common Part Convergence Sublayer
    - Service Specific Convergence Sublayer

At each layer of this model is an interface to Layer Management. In turn, each of the Layer Management entities interfaces to Plane Management. These two management elements form the Management Plane of the B-ISDN PRM. The management plane is the interface that provides services for the management plane impacts two key areas that are visible to the user:

- Network Management
- Traffic Management

Network management activities fall into the category of what is referred to as Operations Administration, Maintenance, and Provisioning (OAM&P). These procedures establish the parameters that control traffic management. Traffic management consists of a set of procedures that monitor and control connection establishment, and the ATM cell traffic carried over those connections. In the next two chapters we will discuss management plane interactions; this chapter will focus on OAM&P, while the next will discuss traffic management.

Figure 5-1 presents a model of the interaction between B-ISDN layers and the management plane [1]. Layer Management (LM) consists of individual functional blocks that perform management functions on

Figure 5-1.
Plane/layer management structure of an ATM network element.

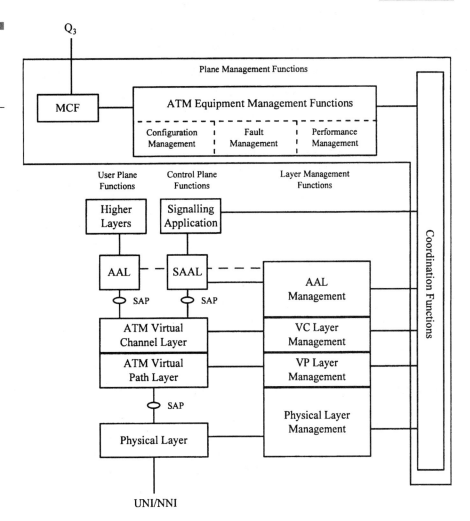

the corresponding sublayer in the User and Control planes. Plane management deals with the set of functions that are applicable to the network element as a whole. Plane management consists of:

- ATM Element Management Functions (AEMF)
- Message Communications Functions (MCF)
- Coordination Functions (CoF)

ATM Element Management Functions support the core OAM&P tasks:

- Configuration Management
- Fault Management
- Performance Management

Message Communications Functions support the exchange of messages between the AEMF and a network management system. This is referred to as the Q3 interface (ITU terminology), or M1, M2, or M4 interface (ATM Forum terminology). The Coordination Functions support the following functions:

1. Interface the AEMF to individual LM blocks; AEMF requests are dispatched to the appropriate LM functional blocks.
2. Provides communication between the Signalling Application and Layer management. When a SETUP request is received by the signalling application it coordinates its response with the management plane to determine if the connection establishment request is to be accepted or denied at that network node.

2. Operations Administration and Maintenance

Operations administration and maintenance is responsible for fault management and performance management functions [2]. OAM performs its functions through the injection of OAM information into the traffic carried over the network. This information is referred to as a *flow*.

2.1 OAM Levels and Flows

Figure 5-2 shows the hierarchical structure of the ATM transport architecture for SONET/SDH. The levels within the hierarchy are significant because of their direct mapping to layer management entities. Individual OAM flows are identified according to the level in the hierarchy at which the flow is terminated. Thus, we have the following OAM flows:

Chapter 5: Operations Administration, Maintenance, and Provisioning

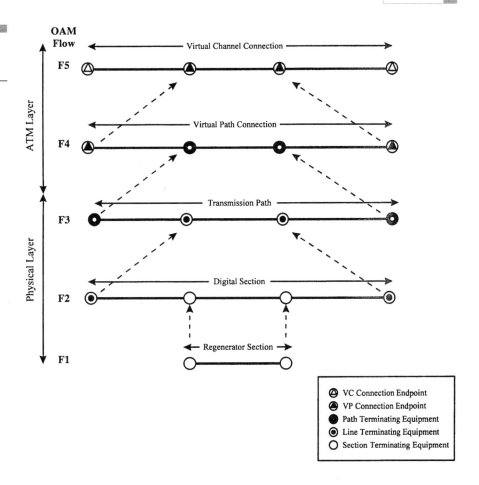

Figure 5-2.
OAM hierarchical structure.

- F1: OAM traffic that originates and terminates within a regenerator section
- F2: OAM traffic over a digital section
- F3: OAM traffic over a transmission path
- F4: OAM traffic over a virtual path connection
- F5: OAM traffic over a virtual channel connection

Not all flows need to be present as the OAM functions of any missing level are performed at the next higher level. Flows F1 through F3, which measure the bit error rate performance of the transmission medium, are managed by Physical Layer management. F4 and F5 flows, which are concerned with the cell delivery performance of an ATM connection, are associated with ATM Layer management.

2.1.1 Physical Layer OAM Flows

At this writing, OAM flows are defined for two different transmission structures [3]:

1. SONET/SDH: ATM cells are transported over a physical medium within a SONET/SDH frame structure.
2. Cell Based: ATM cells are transported with no frame structure

SONET/SDH systems, which allow multiplexing of frame structures (e.g., SONET STS-1 frame structure may be multiplexed into a higher rate STS-3 frame), support F1, F2, and F3 flows.

Cell-based transport systems have no frame structure; the transported entities are individual ATM cells. Consequently, the notion of multiplexing of cell traffic to higher data rates results in no changes in the structure of the transported entities. Cell-based transport supports only F1 and F3 flows.

F1, F2, and F3 information is carried in the frame structure headers for SONET and is extracted from the frame structure during physical layer processing. In cell-based systems, however, OAM cells must be injected directly into the traffic flow to transport physical layer OAM information. These flows are identified by the contents of the ATM cell header as shown in Table 5-1.

During physical layer processing, these cells are identified as containing OAM information, extracted from the cell stream and never seen at the ATM layer. Since these cells are processed at the physical layer, ATM layer header field labels (VPI, VCI, PTI, and CLP) have no meaning. Therefore, Table 1 refers to bit patterns in the first 4 octets of the ATM cell header. Note that the HEC field (contained in Octet 5 of the ATM cell header) must be valid, as HEC verification is a physical layer function.

2.1.2 ATM Layer OAM Flows

F4 and F5 flow information is always carried in ATM cells. These cells are injected into the appropriate VPC or VCC cell stream along with

TABLE 5-1. Header pattern for OAM cell identification

Flow	Octet 1	Octet 2	Octet 3	Octet 4
F1	00000000	00000000	00000000	00000011
F3	00000000	00000000	00000000	00001001

subscriber traffic. Within each flow category there are two types of F4/F5 flows:

1. End-to-End: describes flows that are terminated by VPC (F4) or VCC (F5) connection endpoints
2. Segment: describes flows that travel over a predetermined number of VP or VC links

F4 flows are carried over specific VCCs within a virtual path and are identified by the VCI field in the ATM cell header:

- VCI = 3; for Segment F4
- VCI = 4; for End-to-End F4

While F5 flows are carried over individual VCCs and are distinguished from other (such as user) traffic over the same VCC by the PTI field:

- PTI = 4; for Segment F5
- PTI = 5; for End-to-End F5

2.1.2.1 ATM Layer OAM Functions OAM functions performed at the ATM Layer are:

- Alarm Indication Signal (AIS)—alerts a downstream (receiving) connection endpoint of a detected failure. A failure detected at any level in the OAM hierarchy results in AIS at each of the higher levels in the hierarchy for connections that are impacted by the failure (consequently, there is no F1 AIS).
- Remote Defect Indication (RDI)—notifies the upstream (transmitting) connection endpoint of a failure detected in the downstream direction. RDI cells are sent when an AIS cell is detected.
- Continuity Check—used within a network to provide continuous detection of ATM layer failures. This is equivalent to F1-F3 OAM flows, which provide continuous detection of failures within the physical layer. In the continuity check procedure, a source-point node sends continuity check cells at predetermined times, to a sink-point node which receives them (Figure 5-3). If the sink-point does not receive continuity cells as expected, AIS and RDI are sent.
- Loopback—allows OAM cells to be inserted at a source-point in the network and sent to a destination-point, where the cell is

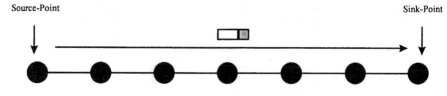

Figure 5-3.
Continuity check OAM cell from source to sink.

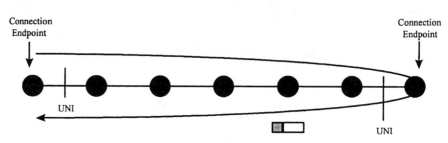

Figure 5-4a.
End-to-end loopback.

looped back in the return direction to the source-point. Loopbacks fall into the following categories (Figure 5-4):

- End-to-end loopback: a loopback cell is inserted at a connection endpoint and looped back at the connection endpoint at the other end (Figure 5-4a).
- Access line loopback: a segment loopback cell (F4 or F5) is inserted by the customer at a connection endpoint and looped back at the first ATM node across the UNI (Figure 5-4b).
- Inter-domain loopback: a segment loopback is inserted by one network operator and looped back at the first ATM node after crossing a B-ICI to an adjacent network operator domain (Figure 5-4c).
- Network-to-endpoint loopback: allows a network operator to insert a segment loopback cell that is looped back at a connection endpoint that is served by another operator domain (Figure 5-4d).
- Intra-domain loopback: a segment loopback cell is inserted at a node within a network operator domain, and looped back at another node within the same domain (Figure 5-4e).
- Performance Monitoring: performs quantitative statistics collection on the performance of the network in delivering cell traffic.

Figure 5-5 shows the OAM Cell format. The OAM Type field identifies the type of management function being performed by the cell (e.g., fault management, performance management, or activation/deactivation of these capabilities over the associated connection), and the Function Type

Chapter 5: Operations Administration, Maintenance, and Provisioning

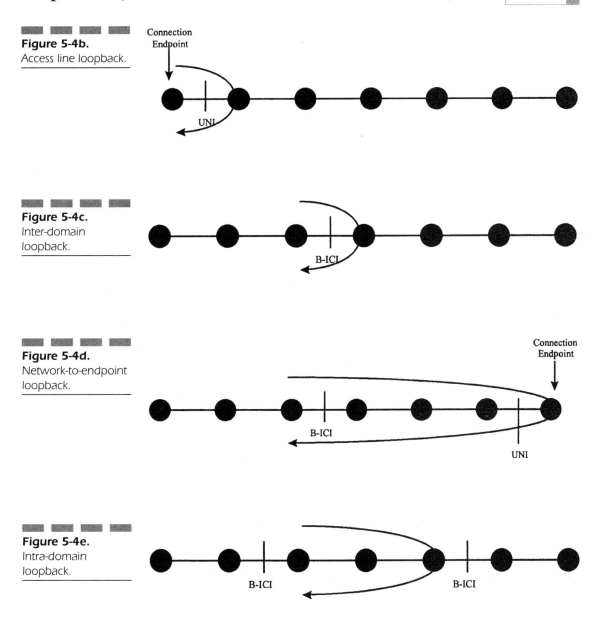

Figure 5-4b. Access line loopback.

Figure 5-4c. Inter-domain loopback.

Figure 5-4d. Network-to-endpoint loopback.

Figure 5-4e. Intra-domain loopback.

Figure 5-5. OAM cell format.

field identifies the actual function being performed. The Function Specific Field contains information relevant to the function being performed.

Figures 5-6 through 5-9 present the formats for the Function Specific Field for each OAM type. Figure 5-6 shows the Function Specific Field for AIS and RDI cells. The Defect Type field, when specified, provides details about the detected failure (e.g., if it was in the VP or VC layer), while the Defect Location field (when provided) contains an identifier (e.g., of an ATM network element) that indicates the location at which the failure was detected.

The Function Specific Field for loopback cells, shown in Figure 5-7, contains the following fields:

- Loopback Indication: provides an indication of whether the cell has reached the destination-point and been looped back.
- Correlation Tag: allows the sender to correlate a transmitted loopback cell with a received cell. This capability allows a source-point to send multiple loopback cells to different (or the same) destination-points.

Figure 5-6. AIS/RDI function specific field.

Defect Type (Optional)	Defect Location (Optional)	Reserved
1 octet	16 octets	28 octets

Figure 5-7. Loopback cell function specific field.

Loopback Indication	Correlation Tag	Loopback Location ID	Source ID (Optional)	Reserved
5 octets	4 octets	16 octets	16 octets	8 octets

Figure 5-8. Forward/backward performance monitoring cell function specific field.

MCSN	TUC_{0+1}	$BEDC_{0+1}$	TUC_0	TSTP (Optional)	Reserved	$TRCC_0$	$BLER_{0+1}$	$TRCC_{0+1}$
8 bits	16 bits	16 bits	16 bits	32 bits	29 octets	16 bits	16 bits	16 bits

Figure 5-9. Activation/deactivation cell function specific field.

Message ID	Directions of Actions	Correlation Tag	PM Block Size A-B	PM Block Size B-A	Reserved
6 bits	2 bits	8 bits	4 bits	4 bits	336 bits

- Loopback Location ID: identifies the destination-point in the network at which the cell is to be looped back.
- Source ID: identifies the source-point of the loopback cell.

There is no function specific information associated with the continuity check function.

The format of the Function Specific Field for forward and backward performance monitoring cells is shown in Figure 5-8:

- Monitoring Cell Sequence Number (MCSN): a sequence number used to correlate a forward performance-monitoring cell with the response sent in the backward direction.
- Total User Cell for CLP_{0+1} (TUC_{0+1}): indicates a current count for the number of user cells (which excludes OAM cells) sent from the source-point with CLP = 0 and CLP = 1. This value is set in the forward performance-monitoring cell and copied into the TUC_{0+1} field in the corresponding backward cell. The difference in TUC_{0+1} values for succeeding forward performance monitoring cells indicates the number of user cells in a performance-monitoring block (the unit of measure in performance monitoring).
- Block Error Detection Code for CLP_{0+1} ($BEDC_{0+1}$): this field, which is used in forward monitoring cells only, carries even parity Bit Interleaved Parity code over 16-bit segments (BIP-16) computed over the payload fields for the entire block of user cells. In BIP-16, the entire block is divided into segments each containing a contiguous group of 16 bits. The least significant bit of the $BEDC_{0+1}$ field represents even parity over the LSB of each segment, etc.
- Total User Cell for CLP_0 (TUC_0): similar to TUC_{0+1} but indicates a current count that includes only the number of user cells sent from the source-point with CLP = 0.
- Time Stamp (TSTP): when provided, this field represents a time stamp for when the performance-monitoring cell was inserted into the ATM cell stream at the source-point.
- Total Received Cell Count for CLP_{0+1} ($TRCC_{0+1}$): this field, which is used in backward monitoring cells only, indicates the number of user cells received at the destination-point with CLP = 0 and CLP = 1. When received at the source-point, the $TRCC_{0+1}$ counts from succeeding backward monitoring cells and the corresponding TUC_{0+1} counts from succeeding forward monitoring cells that may be used for the computation of Cell Loss Ratio for CLP = 0 and CLP = 1 user cell traffic (CLR_{0+1}): a QoS parameter.

- Total Received Cell Count for CLP_0 ($TRCC_0$): similar to $TRCC_{0+1}$ but indicates the number of user cells received at the destination-point with CLP = 0. May be used at the source point for CLR_0 computation.
- Block Error Result for CLP_{0+1} ($BLER_{0+1}$): this field, which is used in backward monitoring cells only, indicates disparities between the BIP-16 computed at the source-point over the transmitted block of user cells and the BIP-16 computed at the destination-point over the received block. This value is only meaningful if there have been no lost user cells over the measured block of user cells.

Figure 5-9 shows the format for the Function Specific Field for OAM cells used to activate or deactivate the OAM function indicated in the Function Type field of the OAM cell:

- Message ID: indicates whether the field contains a command or response:
    - Commands
        - Activate OAM function
        - Deactivate OAM function
    - Responses
        - Activation confirmed
        - Activation denied
        - Deactivation confirmed
        - Deactivation denied
- Directions of Action: indicates whether the OAM function is to be performed in the direction of Point_A to Point_B only, Point_B to Point_A only, or bidirectional between Point_A and Point_B.
- Correlation Tag.
- PM Block Size A-B: used in performance monitoring activation commands to specify the performance monitoring block size in the Point_A to Point_B direction.
- PM Block Size B-A: specifies the performance monitoring block size in the Point_B to Point_A direction.

2.1.2.2 ATM Layer Management Plane Interface Figure 5-10 shows the interaction between the ATM layer management entity and the ATM layer entity [4]. The ATM layer management (ATMM) entity interfaces with the ATM layer entity using the following primitives:

- ATMM-DATA request: which allows the ATM layer management entity to inject ATM cells into the cell stream of an existing connection. The parameters of this primitive include the ATM cell

Chapter 5: Operations Administration, Maintenance, and Provisioning

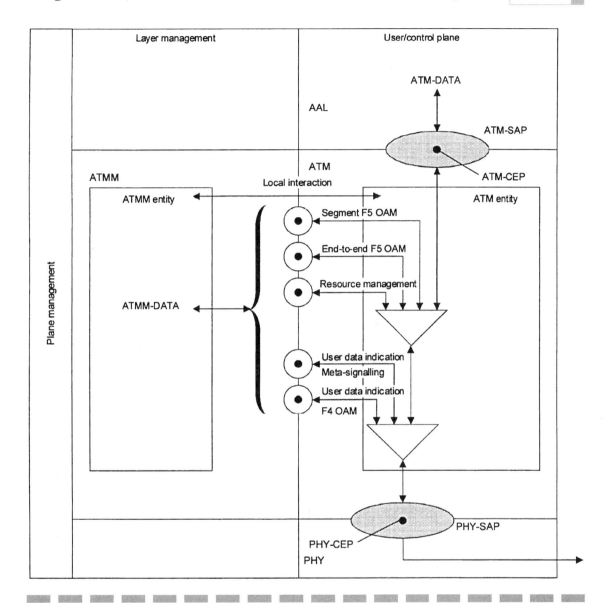

Figure 5-10. Interactions between ATM layer and ATM management layer.

payload, the requested CLP, and a physical layer connection endpoint identifier (PHY-CEI). This last parameter provides the equivalent of the physical layer SAP, and allows ATM layer management to determine characteristics about the interface (such as whether the interface is an NNI or UNI).

- ATMM-DATA indication: signals the arrival of a cell for the management plane. Parameters include the ATM cell payload, the congestion indication, and CLP from the ATM cell header, along with the PHY-CEI of the receiving interface.

In addition to OAM cell transfers across this interface, Figure 5-10 also shows resource management cell traffic at the VC level within the ATM layer. Resource management cells allow ATM connections to dynamically change connection bandwidth. This topic will be discussed in the Traffic Management chapter.

3. Provisioning

The provisioning task interacts with the configuration management function of the AEMF [5]. In addition to administering a number of technology-specific element layer parameters, configuration management involves:

- Interface configuration: establishing the parameters that configure the network element interfaces.
- Bandwidth configuration: allocation of maximum receive and transmit bandwidth available at an interface.
- PVC management: establishment and release of PVC.

Interface configuration establishes the type of interface (UNI, intra-network NNI, or B-ICI), as well as establishing the range of VPI and VCI values that may be used for connections established over the interface. This information, along with bandwidth configuration are used within the traffic management function to determine whether a connection establishment request (either SVC or PVC) is to be accepted or rejected.

PVC establishment is performed from a Network Management System (NMS). A PVC is established by identifying the connection endpoints, and by defining the quality of service, and appropriate traffic descriptors for the connection. Connections may be created, deleted, modified (by changing the connection endpoints or other connection parameters), or queried (to retrieve connection parameters). Establishment of an end-to-end connection may be accomplished in a number of ways, including:

1. Using a Network Layer Management system (as defined by the TMN model) which has a view of an entire network. The NMS will accept the PVC connection establishment request and dispatch

commands to the appropriate Network Element Management Layer entities, which will configure a set of network elements along the path between connection endpoints. An obvious drawback to this approach is that if the connection spans multiple networks, there must be automated procedures based upon the existence of Q and X (ITU) interfaces, or M1-M5 (ATM Forum) interfaces among the various NMS, or manual procedures for coordinating the activities of the operators of each network.

2. Lacking an NMS, using a Network Element Management Layer system to configure each network element individually. This is an unwieldy and error-prone procedure for anything other than the smallest of networks.

3. "Soft" PVC (SPVC) procedures [6]. SPVC was defined by the ATM Forum to provide a mechanism for simplifying PVC establishment. This procedure allows a network element management layer system to enter a PVC establishment request at an individual network node. This request generates a SETUP request, which is subsequently processed within the Control Plane (within the B-ISDN PRM) using PNNI, or B-ISUP procedures.

4. Integrated Local Management Interface

An NMS interacts with the MCF within the Plane Management layer entity through a Q3 or M1, M2, or M4 interface. The TMN and ATM network management models further assume that network management interactions across network boundaries or across a UNI access take place across X, or M3/M5 interfaces. Integrated Local Management Interface (ILMI) provides a means of supporting an M3 functionality across a UNI, or private NNI [7].

ILMI consists of an exchange of messages between user and network side ILMI Management Entities (IMEs) across interfaces such as those shown in Figure 5-11. Each UNI managed by ILMI must have its own ILMI VCC. The obvious drawback to this approach to network management is that the ability to provide remote management of a resource is lost in the event of failure of the interface.

The ATM interface Management Information Base contains managed objects at the Physical Layer, ATM Layer, Virtual Path Connection Layer,

Figure 5-11.
ILMI interfaces.

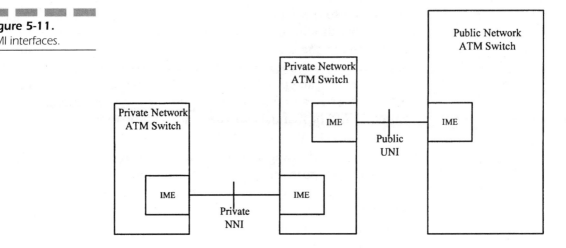

and the Virtual Channel Connection Layer. In addition, the ATM interface MIB supports the following managed objects:

- Address Registration: provides a mechanism for end systems to automatically configure an ATM interface address on the ATM switch across the UNI. This mechanism allows end stations to be added, removed, or moved to another interface without manual configuration at the ATM switch.
- Service Registry: provides information that allows an end system to locate ATM network services. This mechanism is used in support of the LAN Emulation services (LAN Emulation is discussed in Chapter 8) by providing information to end stations on where to locate the LAN Emulation Configuration Server (LECS).

ILMI uses the AAL Type 5 CPCS (with null SSCS). Management plane messages are transferred between peer IMEs by AAL management entities using CPCS-UNITDATA request and indicate primitives.

References

1. Types and General Characteristics of ATM Equipment, International Telecommunication Union, Recommendation I.731, March 1996.
2. B-ISDN Operation and Maintenance Principles and Functions, International Telecommunication Union, Recommendation I.610, November 1995.

3. B-ISDN User-Network Interface—Physical Layer Specification, International Telecommunication Union, Recommendation I.432, March 1993.

4. B-ISDN ATM Layer Specification, International Telecommunication Union, Recommendation I.361, March 1993.

5. Asynchronous Transfer Mode Management of the Network Element View, International Telecommunication Union, Recommendation I.751, March 1996.

6. Private Network-Network Interface Specification Version 1.0 (PNNI 1.0), ATM Forum, Document af-pnni-0055.000, March 1996.

7. Integrated Local Management Interface (ILMI) Specification Version 4.0, ATM Forum, Document af-ilmi-0065.000, September 1996.

CHAPTER 6

Traffic Management

1. Introduction

One of the objectives of ATM technology is to provide support for a wide variety of applications and services based upon the differential treatment of traffic based upon Quality of Service. The primary role of traffic management is to allow networks to achieve network performance objectives which, in turn, allows the network to meet its QoS commitments to the users.

Traffic management represents a significant area of management plane interaction. The following are some of the generic functions performed by traffic management [1]:

- Connection Admission Control (CAC): defined as the set of actions taken during the call establishment phase to determine whether a connection request will be accepted or rejected.
- Dynamic Feedback Mechanisms: which support the exchange of information between the network and end systems to dynamically regulate the allowable rate of information flow on a connection.
- Usage Parameter Control (UPC): consists of the set of actions taken by the network to monitor and control traffic submitted by end users across the UNI. The objective of UPC is to ensure that established connections do not occupy network resources to the extent of affecting the QoS of other connections. When crossing a B-ICI the analogous set of procedures is referred to as Network Parameter Control (NPC).
- Priority Control: allows the network to assign higher priority to some ATM cell traffic based upon the CLP value in the header.
- Traffic Shaping: refers to techniques used to alter the cell transfer rate to achieve a desired modification of traffic characteristics.
- Network Resource Management (NRM): this is a service architecture that allows logical separation of connections based upon characteristics of the connections.
- Congestion Control: defines actions to be taken by the network in the presence of traffic congestion.

In the area of traffic management, there are a number of differences between the ITU and the ATM Forum. Notwithstanding, an understanding of one standard is sufficient for understanding the other so for simplicity, this discussion will focus on the ATM Forum specification.

2. Quality of Service Parameters

The QoS of a connection is based upon the performance of a network (or set of networks) in transporting cell traffic between connection endpoints. This includes all intervening public and private ATM networks. QoS consists of three parameters that are determined at the time of connection establishment:

- Maximum Cell Transfer Delay (CTD)
- Peak Cell Delay Variation (CDV)
- Cell Loss Ratio (CLR)

and three parameters which are not typically specified to vary by connection:

- Cell Error Ratio (CER)
- Severely Errored Cell Block Ratio (SECBR)
- Cell Misinsertion Rate (CMR)

Information about CTD, CDV, and CLR for a connection are estimated within the network and provided to the connection endpoints during call establishment. Such estimates may be derived by a number of means, for example, a formula that computes these parameters based upon the number of nodes, and/or B-ICI transited along the route between connection endpoints. Since many applications are sensitive to these parameters, a connection may be rejected or terminated by either connection endpoint if the estimates returned by the network are unacceptable. A network may also associate quantitative values for QoS parameters with specific QoS classes. Thus at connection establishment, a subscriber may request that service over the connection be provided at a level which is predetermined for a specified QoS class.

Figure 6-1 describes pictorially the concepts behind CTD, CDV, and CLR. The maximum CTD is based upon the $(1 - \alpha)$ quintile of a cell transfer delay probability distribution function. There is assumed to be some minimum time required to transfer cells between the connection endpoints (Fixed Delay). The maximum CTD is assumed to be the maximum CTD acceptable to the subscribers of the given service. Cells not delivered within the maximum CTD are considered lost (α% of total user cell traffic), at least in principle. In reality, the ATM layer has no mechanism for monitoring late arrival of individual user cells; there is only the assumption that a cell not delivered within a given time will never

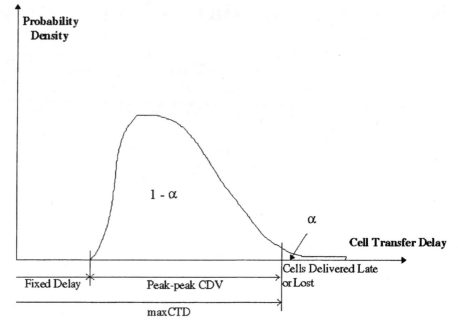

Figure 6-1.
Probability density model for cell transfer delay.

arrive (which can be detected by performance monitoring procedures). Peak CDV is defined by:

$$\text{Peak CDV} = \text{Max CTD} - \text{Fixed Delay}$$

The definitions for CLR is:

$$\text{CLR} = \frac{\text{Lost Cells}}{\text{Total Transmitted Cells}}$$

where "lost cell" is a cell that does not arrive within the maximum CTD interval.

CER is defined as:

$$\text{CER} = \frac{\text{Errored Cells}}{\text{Successfully Transferred Cells} + \text{Errored Cells}}$$

where an "errored cell" is one with an uncorrectable error in the cell header, or a cell with a corrupted cell payload. A "successfully transferred cell" is one transferred with no error.

Chapter 6: Traffic Management

The equations defining the SECBR parameter is:

$$\text{SECBR} = \frac{\text{Severely Errored Cell Blocks}}{\text{Total Transmitted Cell Blocks}}$$

A "severely errored cell block" is a performance-monitoring block in which there has been an unacceptably high number of errored cells [2].

The CMR equation is:

$$\text{CMR} = \frac{\text{Misinserted Cells}}{\text{Measurement Time Interval}}$$

A misinserted cell is a cell that is received at a connection endpoint, which was not transmitted by the peer connection endpoint. Cell misinsertions are typically the result of an undetected error in the ATM cell header, which results in the cell being delivered to the wrong destination. Since misinserted cells are uncorrelated to any cell traffic actually transmitted over the connection under observation, CMR is defined as a rate per unit time, rather than as a ratio.

2.1 Sources of Delay

From the application perspective all sources of delay must be considered. These include:

1. Information coding and decoding delay

2. AAL segmentation and reassembly delay

3. End-to-End cell transfer delay, consisting of:

- Total inter-ATM node transmission delay
- Total ATM node processing delay due to cell queuing, input to output transit delay, etc.

QoS, within the ATM context, delay is based upon (3) only. Thus, if there are intermediate nodes (between end subscribers) which provide interworking functions (such as a B-ISDN subscriber communicating with a Frame Relay subscriber), QoS commitments are made only over the contiguous portion of the route between the entry point to the B-ISDN and the exit point.

2.2 Sources of CDV

ATM layer functions may alter the traffic characteristics of ATM connections by introducing cell delay variation. Examples of sources of CDV are:

1. Multiplexing of cell traffic from different connections over the same transmission path
2. Physical layer overhead insertion
3. OAM cell insertion

Figure 6-2 shows the impact of a combination of CBR traffic, VBR traffic, OAM traffic, and physical layer overhead on CDV for connections carried over the interface. In this example an ATM TE multiplexes traffic from three different connections over the same physical medium. Connection_A is CBR, Connection_B is ABR, and Connection_C is VBR. Interspersed in with the user traffic is an OAM cell injected at the ATM layer from ATM layer management over Connection_A, and a resource management cell over Connection_B. Overhead is also injected into the bit stream at the physical layer. The net result shows that the timing relation between cells on an individual connection as they were submitted at the ATM layer varies considerably from the timing relation as seen at the physical layer. For example, at the physical layer, Connection_A appears to be VBR in character.

2.3 Measurement of QoS Parameters

CLR and SECBR may be directly measured using performance monitoring OAM cells. CER is most reliably measured by establishment of a test VCC over which is transmitted a series of ATM cells with known payloads. A test set at the receiving end of the connection is then able to monitor the received cell stream for errors. CMR may be measured by establishing a test connection over which no traffic is transmitted. Any cell traffic received at the receiving side test set will then be recorded as a misinserted cell. The CMR may then be determined by dividing the number of misinserted cells by the measurement time interval.

CDV may be measured by one of two procedures:

Chapter 6: Traffic Management

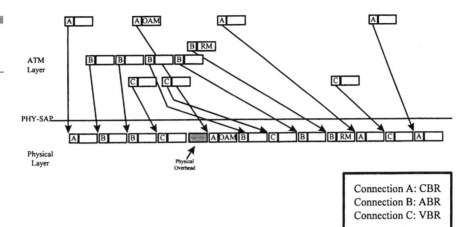

Figure 6-2.
Origins of cell delay variation.

Connection A: CBR
Connection B: ABR
Connection C: VBR

1. Two Point CDV—using time-stamped performance monitoring OAM cells, which produces a series of CTD observations between measurement points.

2. One Point CDV—: approximates two point CDV based upon a series of observations at a single measurement point. This method assumes a CBR source from which is computed a reference arrival time. The expected arrival time of the kth cell at the measurement point, c(k) is compared to a(k), the actual arrival time of the kth cell. The variable, Q(k) is the CDV accumulated over the k cell arrivals. For y(k) < 0:

$$Q(k) = Q(k-1) - y(k)$$

for y(k) > 0:

$$Q(k) = \max(Q(k-1), y(k))$$

The difference, y(k), is the CDV for the kth cell arrival. The series of y(k) values may be observed directly in AAL Type 1 cells that use the Residual Time Stamp Method.

2.4 Factors Impacting QoS

The following are factors that may result in QoS degradation:

1. Propagation Delay: caused by the physical media that transport the bits comprising ATM cells between connection endpoints. Propagation delay is dependent upon distance.

2. Random error and/or error bursts that are sometimes introduced on physical media due to factors such as EMI, &c.
3. Switch Architecture: the design of an ATM switch can have a significant impact on performance. Some aspects are:

   - whether the switch fabric design is blocking or nonblocking (i.e., whether there is or is not always a path between any input port and any output port).
   - the strategy for allocation of buffer capacity among switch ports; buffer capacity may be dedicated to specific ports, or buffer capacity may form a pool which is shared among any buffers needing cell queuing capacity.
   - priority queuing mechanisms: QoS within a switch may be implemented by associating multiple queues with an output port. Cells in the various queues then receive different input or output priorities. The scheme for servicing these queues has a direct impact on QoS.

4. Buffer Capacity available at an individual port, measured in units of cells; the greater the capacity, the less likelihood of congestion under heavy traffic loads and resultant cell discard.
5. Traffic Load: this is the aggregate cell rate of all connections sharing a common route on the network. Congestion is seen at output ports when the aggregate rate of switched traffic being directed to the port exceeds the bit rate of the transmission media for extended periods of time.
6. Number of Tandem Nodes: this is the number of ATM switching nodes that are traversed for a particular connection.
7. Failures: these can be either due to facilities failures, equipment failures, or software failures. Failures may result in network resources becoming unavailable, which in turn negatively impacts connections which are dependent upon those resources.

3. Traffic Parameters

Traffic parameters describe the inherent characteristics of the traffic source. The set of traffic parameters that characterize the traffic source is referred to as a *traffic descriptor*. Along with the QoS parameters, the traffic descriptor defines a *traffic contract*. This traffic contract determines the category of service provided at the ATM layer and represents a commitment from the network to the end user. The traffic parameters discussed in this section are:

Chapter 6: Traffic Management

- Peak Cell Rate (PCR)
- Sustainable Cell Rate (SCR)
- Maximum Burst Size (MBS)
- Minimum Cell Rate (MCR)

3.1 Peak Cell Rate

The peak cell rate specifies an upper bound on the rate (in cells/sec) at which cell traffic can be submitted on an ATM connection. The location at which the PCR (as well as the other traffic parameters discussed in this section) is observed is at the physical layer SAP. The PCR is encoded as a 24-bit value that supports a range from 1 cell/sec to 16,777,216 cells/sec.

3.2 Sustainable Cell Rate

The sustainable cell rate defines an upper bound on the rate at which cells may be submitted on an ATM connection over a time scale that is long compared to the time scale for which cells may submitted at the PCR. SCR is also encoded as a 24-bit value.

3.3 Maximum Burst Size

If an SCR is defined for a connection, the maximum burst size defines the maximum number of consecutive cells that may be submitted at the PCR. Burst tolerance, which is based upon MBS, PCR, and SCR, represents the time scale over which cells may be continuously submitted at the PCR. MBS is a dimensionless quantity defined as a 24-bit integer.

3.4 Minimum Cell Rate

The minimum cell rate, which is defined as a 24-bit integer at connection establishment, defines a cell submission rate that will always be guaranteed over the ATM connection. For connections employing an

MCR specification, there is a negotiation procedure between the endpoints and the network for requesting higher bandwidth allocations (up to the PCR) on the connection. The information for these negotiations is carried within Resource Management cells, which are processed by the ATM layer management entity.

4. ATM Service Categories

The ATM service category request for a connection is negotiated at the time of call establishment in the SETUP message. Table 6-1 presents descriptions of each of the five ATM layer service categories as defined by the ATM Forum:

1. Constant Bit Rate (CBR): the CBR category used by connections that request that a fixed amount of bandwidth is continuously available for the duration of the connection. This amount of bandwidth is characterized by the PCR value. The subscriber receives a QoS commitment from the network for transmitted cell traffic sent at the PCR. Traffic sent at rates above the PCR (nonconforming traffic) does not receive a committed QoS. The subscriber may transmit at any rate below the negotiated PCR at any time (including periods where no traffic is sent) for any duration without impacting the QoS commitments for this traffic or any subsequent traffic transmitted at the PCR. CBR service is intended for real-time applications requiring tightly constrained cell delay variation (such as voice, video, and circuit emulation) but may be used for any application.

2. Real-Time Variable Bit Rate (rt-VBR): this category is intended for real-time applications over connections with variable bit rates over the lifetime of the connection. rt-VBR connections are characterized by PCR, SCR, and MBS parameters. Traffic sources can be expected to transmit cells at varying rates within the parameters defined for the connection and receive a committed QoS from the network. Nonconforming traffic does not receive committed QoS.

3. Non-Real-Time Variable Bit Rate (nrt-VBR): the non-real-time VBR category is intended for VBR traffic which does not require constraints on cell delay and cell delay variation.

TABLE 6-1.

ATM Service category attributes

		ATM Layer Service Category				
Attribute	**CBR**	**rt-VBR**	**nrt-VBR**	**UBR**	**ABR**	
PCR	specified	specified	specified	optional	specified	
SCR, MBS	n/a	specified	specified	n/a	n/a	
MCR	n/a	n/a	n/a	n/a	specified	
maximum CTD	specified	specified	unspecified	unspecified	unspecified	
CDV	specified	specified	unspecified	unspecified	unspecified	
CLR	specified	specified	specified	unspecified	optional	
RM Cell Rate	n/a	n/a	n/a	n/a	specified	

4. Unspecified Bit Rate (UBR): UBR is intended for non-real-time applications such as traditional computer communications (file transfer, electronic mail, etc.). No traffic related service guarantees are granted with respect to PCR, SCR, or MBS. There are also no CLR or cell delay commitments made in UBR service. In essence, UBR is a "take whatever bandwidth you can get" level of service.

5. Available Bit Rate (ABR): ABR is another non-real-time service category. In ABR service, a subscriber requests an MCR which establishes a minimum bandwidth for which the user may expect low CLR for the duration of the connection. Whether a specific CLR is specified is a network option. Resource management cells allow connection endpoints to request higher (or lower) cell transmission rates (up to the PCR) according to the temporal needs of the application. As long as the cell transmission rate is within the current allowable bit rate granted by the network, the subscriber can expect low CLR. The subscriber may also expect to receive a fair (according to network policies) allocation of available network bandwidth in response to bandwidth requests. Once granted, however, the network may later indicate to the subscriber (through resource management cells) that previously granted bandwidth has been rescinded. At no point, however, is the subscriber to be required by the network to reduce the cell transmission rate below the MCR.

Selection of one service category over another by a subscriber may be influenced by factors that may include:

1. The nature of the traffic source.
2. The services available by a serving network operator.
3. The relative cost to the subscriber of service category offerings.

4.1 Available Bit Rate Service

ABR is a flow-controlled ATM layer service in which the traffic source adapts its cell transmission rate over a connection to changing conditions within the network. The objective of ABR is to provide access to unused network bandwidth (up to the PCR) when available. The information necessary to perform these rate adaptations is communicated through special control cells known as Resource Management (RM) cells. The flow control negotiations take place between:

1. A connection endpoint sending an RM cell (source) and the network, and;

2. A source and the connection endpoint that ultimately receives the RM cell (destination) over a control loop as shown in Figure 6-3. In ABR, two connection endpoints must be served by a bidirectional connection; therefore, each endpoint is able to serve both sender and destination roles over separate control loops. For simplicity Figure 6-3 shows only one such loop. At this writing, there is no specification for ABR on point-to-multipoint connections.

As can be seen in Figure 6-3a, the RM control loop is similar to an end-to-end F5 flow (for VPC RM cells, the VPC must connect a source and destination pair) in which the forward direction is from source to destination, where the RM cell is looped back and sent in the reverse direction back to the source. With RM cells, however, each network node along the path inspects the RM cell and may modify its contents. Also shown, in Figure 6-3b, a network node (or destination) may generate a reverse RM cell that is sent to the source. These RM cells are typically sent to instruct the source to reduce the current cell transmission rate.

At connection establishment, the following parameters are established relative to ABR service:

- Peak Cell Rate
- Minimum Cell Rate
- Initial Cell Rate (ICR): the rate at which the source may transmit cells over the connection prior to sending of the first RM cell. Defined as a 24-bit integer.
- Transient Buffer Exposure (TBE): this value determines the number of cells that the source will be allowed to transmit prior to being subject to rate adjustment. Also defined as a 24-bit integer.

Chapter 6: Traffic Management

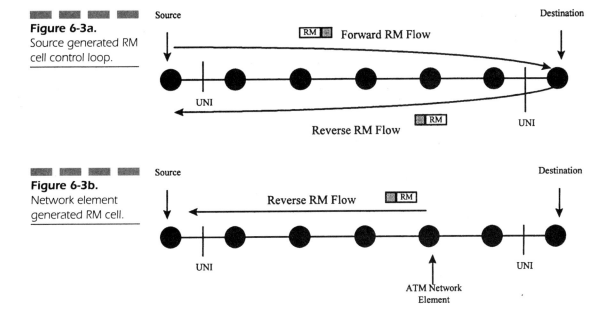

Figure 6-3a. Source generated RM cell control loop.

Figure 6-3b. Network element generated RM cell.

4.1.1 RM Cell Format

Figure 6-4 shows the format of the RM cell, which contains the following fields:

- Direction (DIR): identifies the direction (forward = 1/backward = 0) of the RM cell.

- Backward Error Congestion Notification (BECN): An intermediate switch node or destination may generate a reverse RM cell without having received a forward RM cell from the source. BECN = 1 distinguishes a reverse RM cell generated by the destination, or an intermediate switch node from one sent in response to a forward cell originally generated by the source. BECN is a notification to the source that traffic congestion conditions exist within the network and that the source may need to adapt its cell transmission rate as indicated in the RM cell. A destination may generate BECN to the source in response to receiving ATM cells with EFCI = 1 which indicates that the cell encountered network congestion in the source to destination path.

- Congestion Indication (CI): CI = 1 also indicates congestion within the network but is communicated to the source in a reverse RM

Figure 6-4.
Resource management cell format.

BIT								OCTET
8	7	6	5	4	3	2	1	
ATM Cell Header								1-5
Protocol Identifier								6
DIR	BN	CI	NI	RESERVED				7
ER								8
								9
CCR								10
								11
MCR								12
								13
RESERVED								14-51
CRC								52-53

cell which was generated in response to a forward RM cell originated by the source.

- No Increase (NI): NI = 1 indicates to the source that it is not allowed to increase the cell rate for the connection. This does not necessarily indicate that a cell rate decrease is required. For example, if CI = 0 the network may set this indication to instruct the source not to request additional bandwidth. This might be done preemptively to prevent congestion from occurring within the network. A source may send a forward RM cell with NI = 1 to confirm the current cell rate and to indicate that no increase is requested. A source indicates NI = 0 when it wishes to request a cell rate increase.

- Explicit Rate (ER): is used to specify the allowable cell rate (ACR) for cell transmission on the connection. Within RM cells, all cell rates are specified as a 9-bit mantissa (k_{ACR}) and a 5-bit exponent (m_{ACR}) and computed based upon the following equation:

$$ACR = 2^{m_{ACR}} \times \left(1 + \frac{k_{ACR}}{512}\right)$$

Chapter 6: Traffic Management

$$0 \leq m_{ACR} \leq 31$$

$$0 \leq k_{ACR} \leq 511$$

In a source generated RM cell, the source enters values to indicate the requested ACR (up to the PCR) in the forward cell. Any network element along the path may revise this value downward to a value sustainable by that element. This value is not modified by the destination. A network node may also set a value for ER along with CI = 1 and NI = 1 to indicate to the source that it is to reduce its cell rate.

- Current Cell Rate (CCR): is set by the source in forward RM cells to indicate to the network the current ACR as seen by the source. The network may use this information in cell rate assignment decisions. CCR (as are all cell rates within an RM cell) is specified using the same formula as is used to describe ACR. Non-source generated RM cells set this value to 0.

- MCR: set by the source in forward RM cells to the value established at connection establishment. Like CCR, the network has the option of using this information in cell rate assignment decisions. Set to 0 in non-source generated RM cells.

5. Usage Parameter Control

To the degree that the actual traffic from a source does not exceed the bounds set by the traffic parameters for the connection, the traffic pattern is said to be compliant with the traffic contract. The subscriber, under those circumstances may expect that submitted traffic will receive the committed QoS. Traffic that is noncompliant, however, may not receive a committed QoS. The role of Usage Parameter Control (UPC) is to monitor submitted traffic from a source, determine its compliance with the traffic contract, and take appropriate actions. As shown in Figure 6-5, UPC is enforced at the point of network access on the UNI.

Between nodes within a network, compliance verification may not be performed. Since the traffic contract is with the end subscriber, it is

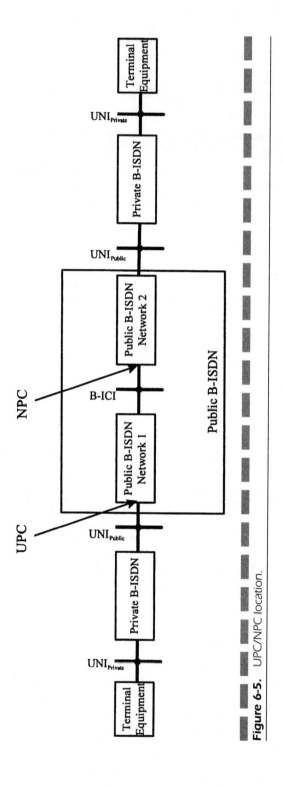

Figure 6-5. UPC/NPC location.

expected that as long as the subscriber traffic is well behaved according to UPC, it is the network operator's responsibility to properly engineer the network to provide the committed QoS to that traffic. If compliance verification were performed between nodes within a network such that traffic were discarded as a result, the network operator would have failed in the commitment to the end subscriber.

Compliance verification may be performed at the B-ICI. This is referred to as Network Parameter Control (NPC). A network operator has a responsibility to ensure that no traffic entering the network, from any source, will negatively affect the QoS of other connections. For instance, excessive CDV introduced from another network could produce "clumping," a temporary burst of cells, which could violate the traffic contract for subscriber connections. The receiving network must take action to ensure that excessively large cell bursts are not allowed into the network.

UPC/NPC is also responsible for verifying that incoming traffic is valid, i.e., associated with an active VCC or VPC, and admissible to the network.

5.1 Generic Cell Rate Algorithm (GCRA)

The specific algorithm for UPC used at any network node is an implementation matter that is not subject to dictate by standards. In any event, the algorithm for UPC/NPC must ensure that compliant traffic receives the committed QoS. The conceptual basis for defining traffic conformance is the Generic Cell Rate Algorithm (GCRA). For each cell arrival (measured at the physical layer SAP), the GCRA determines whether the cell conforms to the traffic contract of the connection. The GCRA expresses the relationship between traffic parameters and CDV. This relationship, shown in Figure 6-6, may be expressed by either the Virtual Scheduling algorithm, or the Leaky Bucket algorithm.

5.1.1 Virtual Scheduling Algorithm

In the virtual scheduling algorithm, the actual time of arrival for the kth cell, $t_a(k)$, is compared against the Theoretical Arrival Time, TAT(k),

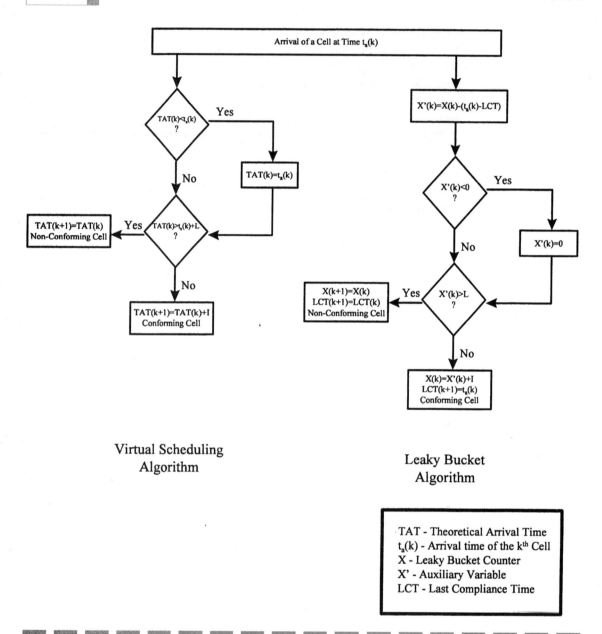

Figure 6-6. Equivalent versions of the generic cell rate algorithm.

which is determined based upon the traffic contract. If $t_a(k) \geq TAT(k)$, the cell has not arrived early. The cell is, therefore, considered conformant. In that case, the algorithm sets then sets TAT(k) equal to ta(k) and computes the expected arrival time for the $(k+1)^{th}$ cell, TAT(k+1):

$$TAT(k + 1) = TAT(k) + I$$

where *I* is the interval between cell arrivals at the rate defined by the traffic contract. From the GCRA perspective, a late arriving cell may be due to the traffic source temporarily reducing the cell rate, which is permissible. On any account, late cells suggest an instantaneous cell rate that is not in excess of that specified by the traffic contract.

If, $t_a(k) < TAT(k)$, the kth cell has arrived earlier than expected. In this case the comparison:

$$TAT(k) > t_a(k) + L$$

determines whether the cell is compliant. The variable *L* indicates a limiting value on how early a cell may arrive and still be compliant. If the cell is compliant, TAT (k + 1) is updated to reflect the expected time of arrival of the $(k + 1)^{th}$ cell based upon the equation TAT (k + 1) = TAT (k) + I. If the cell is not compliant, however, TAT (k + 1) = TAT (k) indicating that the conformance of the $(k + 1)^{th}$ cell will be evaluated against the expected arrival time of the kth cell.

The boundary condition for the first cell arrival in the virtual scheduling algorithm is:

$$TAT(1) = t_a(1)$$

Thus, the first cell can never be noncompliant.

5.1.2 Leaky Bucket Algorithm

The leaky bucket algorithm is an equivalent GCRA that can be viewed as a finite capacity leaky bucket. The "fill level" of the leaky bucket decreases at a continuous rate of 1 unit per unit of time, and "refilled" with *I* additional units for each conforming cell that arrives. The variable X(k) serves as the leaky bucket fill level after the arrival of the $(k-1)^{th}$ cell in the following equation:

$$X'(k) = X(k) - (t_a(k) - LCT(k))$$

LCT is the arrival time of the last compliant cell, and X'(k) is an auxiliary variable that marks the fill level of the leaky bucket at the arrival time of the kth cell. The difference between X(k) and X'(k) represents the amount of leaky bucket drain between cell arrivals. This difference is represented by the second term in the equation, which indicates the amount of time between the *LCT* and the currently arriving cell. The earlier the arrival of the kth cell, the smaller this term will be. In turn, this will make X'(k) larger. If X'(k) ≤ ∅, then the "drain" rate of the leaky bucket is greater than, or equal to the fill rate. In this case, the interval between consecutive cell arrivals is within acceptable limits, and the arriving cell is considered compliant with the traffic contract; the fill level of the leaky bucket for the $(k+1)^{th}$ cell arrival is increased by *I* units, and the LCT (k + 1) is set equal to the time of arrival for the kth cell.

The leaky bucket refill level is never below *I* after arrival of a compliant cell. Traffic conformance operates based upon a "use it or lose it" principle; without a lower bound on the leaky bucket refill level, a very late arriving cell could be followed by a transient burst of cells, which could overwhelm network resources and negatively impact QoS rendered to other connections.

If X'(k) > ∅ a series of cell arrivals has occurred at an aggregate rate such that the leaky bucket fill rate exceeds the drain rate. If the kth cell has arrived so early that the accumulated effect of fills and drains is such that X'(k) > L the arriving cell is declared nonconforming and there is no leaky bucket refill. In this event, the leaky bucket continues to drain in anticipation of the next cell arrival. Thus, we see that the leaky bucket algorithm prevents "overflow" of the leaky bucket.

The boundary conditions for the first cell arrival in the leaky bucket algorithm are:

$$X(1) = 0$$

$$LCT(1) = t_a(1)$$

5.1.3 GCRA Example

Presented below is an example that shows the operation of the virtual scheduling and leaky bucket algorithms. The conditions for the example are:

$$I = 4.5$$

$$L = 1.5$$

which may be expressed in notation as GCRA(4.5, 1.5). As Table 6-2 shows, both algorithms produce the same result. When the traffic pattern consists of a series of early arriving cells, eventually a nonconforming cell is identified at the fifth cell arrival. The next cell, even though it arrives at the same intercell interval as the previous ones, is considered conforming. This cell is followed by a late arriving cell (k = 6) and subsequently by a series of cells which arrive at the expected time.

In terms of traffic contract parameters, I represents the cell transmission rate, while L represents CDV tolerance (CDVT).

5.1.3.1 CDV versus CDVT It should be noted that CDVT is not the same as CDV. CDV is of end-to-end significance while CDVT is of local UNI significance. CDVT is typically set by a network service provider based upon the amount of burst tolerance that the service provider will allow for traffic entering the network over the UNI. There are a number of sources of CDV at a terminal as discussed earlier in this chapter, the CDVT is set by the network operator to allow for bursts of traffic at a rate faster than the contracted rate from the subscriber due to these sources.

TABLE 6-2. Leaky bucket versus virtual scheduling algorithm comparison

	Virtual Scheduling		Leaky Bucket			
k	ta(k)	TAT(k)	ta(k) + L	X'(k)	X(k)	LCT(k)	Conformance
1	0.0	0.0	1.5	0	0	0	compliant
2	4.0	4.5	5.5	0.5	4.5	0.0	compliant
3	8.0	9.0	9.5	1.0	5.0	4.0	compliant
4	12.0	13.5	13.5	1.5	5.5	8.0	compliant
5	16.0	18.0	17.5	2.0	6.0	12.0	noncompliant
6	20.0	18.0	21.5	-2.0	6.0	12.0	compliant
7	28.0	24.5	29.5	-3.5	4.5	20.0	compliant
8	32.5	32.5	34.0	0	4.5	28.0	compliant
9	37.0	37.0	38.5	0	4.5	32.5	compliant
10	41.5	41.5	43.0	0	4.5	37.0	compliant

Notes: I = 4.5, L = 1.5

CDVT may also be used at the B-ICI to allow a network to protect itself from unexpected bursts of traffic from other networks. CDVT, however, is typically not used for traffic policing between switches within a carrier network—the objective of UPC/NPC is to determine that a subscriber (or other network provider) is providing a compliant traffic flow into the network.

5.2 Traffic Contract and Conformance Definitions

The preceding section presented the basic model and concepts of the GCRA. In practice, there may be multiple instances of the GCRA (with different values for I and L) for different flows according to CLP:

- CLP = 0 flow: when present, this is the more stringent GCRA, which is applied to cells with CLP = 0 only.
- Aggregate CLP = 0 and CLP = 1 flows: this GCRA is applied to the total flow of cells from the connection endpoint; independent of the CLP value in the ATM cell header.

Each of these instances are applied to a different element of the traffic contract.

At the ATM cell level, UPC/NPC may take one of the following actions on an arriving cell:

1. Admit the cell into the network.
2. Tag the cell—by changing the CLP value of the cell from CLP = 0 to CLP = 1.
3. Discard the cell—which eliminates it from the cell traffic flow and increases the CLR for the connection.

Admission is granted to cells that are identified as being compliant. Cell tagging is optionally performed in instances where there is more than one instance of the GCRA, on cells that are identified as nonconforming to one element of the traffic contract. Cell discard is performed on cells that are identified by UPC/NPC as being noncompliant with at least one element of the traffic contract.

5.2.1 Conformance Definition for CBR Service

The following is a conformance definition for a source traffic descriptor defined at the Public UNI:

$$GCRA(T_{PCR,0+1}, CDVT_{PCR})$$

where $T_{PCR,0+1}$, is the cell arrival interval for a traffic source, which generates cells at the PCR—defined for the aggregate CLP = 0, and CLP = 1 flow. There is no tagging option for this specification and there is no separate GCRA for the CLP = 0 flow. CBR is based upon fixed bandwidth allocation with no variation regardless of CLP value. CLR is computed based upon the aggregate CLP = 0 + 1 cell traffic.

The CDV tolerance in the specification takes into account the accumulated result of any sources of CDV within the B-TE, and within the Private ATM network (if present).

5.2.2 Conformance Definition for rt-VBR Service

VBR requires a GCRA for the PCR as well as one for the SCR. Since VBR supports bursts of cell traffic at the PCR, this creates a need for additional CDVT in the SCR traffic description, referred to as *burst tolerance*. Given an MBS value, the burst tolerance, BT, is not uniquely defined, but may have any value within the half-closed interval:

$$(MBS - 1) \times (T_{SCR} - T_{PCR}) \le BT < MBS \times (T_{SCR} - T_{PCR})$$

which defines the time scale over which cells may be transmitted continuously at the PCR. By convention, the minimum value is chosen for BT when used in a conformance definition. There are three possible ways to define the conformance definition for this service:

1. GCRA for the PCR(CLP = 0 + 1) traffic descriptor, and GCRA for the SCR(CLP = 0 + 1) traffic descriptor: UPC makes no distinction between CLP = 0 and CLP = 1 traffic. CLR is computed on the aggregate cell traffic.
2. GCRA for the PCR(CLP = 0 + 1) and GCRA for the SCR(CLP = 0) with no cell tagging: UPC enforces conformance on CLP = 0 only

for SCR. UPC enforces conformance on CLP = 0 + 1 for the PCR traffic descriptor. A CLP=0 cell must be compliant with both GCRA definitions while CLP = 1 cells must only be compliant with the PCR GCRA definition. CLR is computed on CLP = 0 traffic.

3. GCRA for the PCR(CLP = 0 + 1), and GCRA for the SCR(CLP = 0) with cell tagging: In this case, noncompliant CLP = 0 cell traffic is tagged and subsequently becomes part of the CLP = 1 flow.

The following are the conformance definitions for each case:
Case 1:

$$GCRA(T_{PCR, 0+1}, CDVT)$$

$$GCRA(T_{SCR, 0+1}, BT + CDVT)$$

Cases 2 and 3:

$$GCRA(T_{PCR, 0+1}, CDVT)$$

$$GCRA(T_{SCR, 0}, BT + CDVT)$$

Burst tolerance has the effect of producing a larger effective CDV tolerance. The difference between cases 2 and 3 above is in how the state variables are updated. In case 2, a CLP-0 cell which is noncompliant to the SCR GCRA definition results in no update of algorithm state variables. In case 3, this cell would be tagged. If it is then found to be compliant with the PCR GCRA definition, the state variables of that definition are updated.

5.2.3 Conformance Definition for nrt-VBR Service

rt-VBR and nrt-VBR are distinguished by the QoS class and/or QoS parameters, but also by the size of the MBS. Since real-time traffic requires more tightly controlled CDV, rt-VBR should be expected to specify smaller MBS than would be the case in an nrt-VBR traffic contract. The conformance definitions are otherwise the same between the two VBR services.

5.2.4 Conformance Definition for UBR Service

There is no conformance definition for UBR service.

5.2.5 Conformance Definition for ABR Service

1. The conformance definition:

$$GCRA(T_{MCR,0}\tau_1)$$

is true for all CLP=0 cells, where:

$$\tau_1 \geq max(t_1) - min(t_1)$$

and the variable t_1 denotes the time from cell transmission at the source to receipt at the destination.

2. A CLP = 0 cell will be declared nonconforming (based upon the one-point CDV observed at the destination) if the arrival times between the cell and preceding cells could not have resulted from Ideal Transmission Times (ITT) of an ABR source. The transmission time of the kth cell is called an ITT if the difference between the transmission time of that cell and that of the $(k-1)^{th}$ cell is greater than or equal to the minimum of:

- the cell arrival interval of the ACR in effect immediately after the transmission time of the $(k-1)^{th}$ cell, and;
- the cell arrival interval of the ACR in effect immediately before the transmission time of the kth cell

The minimum of the two quantities is chosen because ABR allows the traffic source to adapt its cell transmission rate (e.g. to a higher ACR) in response to received RM cells.

3. If we take t_2 to be the sum of:

- the time interval from the departure of a backward RM cell from the destination to the receipt of that cell at the source, and;
- the time interval from the departure of the next cell from the source to the receipt of that cell at the destination.
  and if we set upper and lower bounds for t_2:

$$\tau_3 \leq t_2 \leq \tau_2$$

then we may assume that the cell transmission time of a compliant k^{th} cell will account for ACR modifications indicated from backward RM cells sent from the destination at least τ_2 units of time before transmission of the $(k-1)^{th}$ cell, but no more than τ_3 units of time before transmission of the $(k-1)^{th}$ cell.

5.3 Measurements Associated with UPC/NPC

Measurement data is stored at each network node, which provides a record of the activities of UPC/NPC at a network node. This data may be automatically reported, or retrieved on-demand for presentation at an NMS [3]. These maintained counts include:

1. Discarded cells with CLP = 0 + 1. The count of cells discarded due to UPC/NPC with CLP = 0 or CLP = 1.
2. Discarded cells with CLP=0. The count of cells discarded due to UPC/NPC with CLP = 0.
3. Successfully passed cells with CLP = 0 + 1. The count of cells that have been passed (i.e., not discarded) by UPC/NPC with CLP = 0 or CLP = 1.
4. Successfully passed cells with CLP = 0. The count of cells that have been passed by UPC/NPC with CLP = 0.
5. Count of cells tagged by UPC/NPC. The count of cells with CLP = 0 which were tagged by UPC/ NPC.

6. Connection Admission Control

From each connection establishment request, the CAC function will be able to derive the following information from the traffic contract:

- Values of parameters in the source traffic descriptor.
- Requested and acceptable values of each QoS parameter and the requested QoS class.
- Value of the CDVT.
- Requested conformance definition.

The CAC function makes use of this information to determine:

- Whether the connection can be accepted or not.

- The connection traffic parameters needed by UPC/NPC.
- Allocation of network resources for the connection. This includes VPI/VCI assignments and bandwidth allocations.

7. Network Resource Management

In NRM, virtual paths are an important component of traffic control and resource management in ATM networks. In this context, VPCs provide the following benefits:

- Simplified CAC—VPCs may be preestablished between ATM network nodes. Subsequent established VCCs that traverse these nodes need only be cross-connected to a VCC within a preestablished VPC.
- Implementation of priority control by segregating groups of VCCs according to service category—VPCs may be preestablished between network nodes, with each VPC being assigned to carry traffic of a designated service category. Statistical multiplexing within a VPC, where the aggregate peak of all VCCs may exceed the VPC capacity, is only possible when all VCCs within the VPC can tolerate the QoS that results from statistical multiplexing (such as VBR, ABR, and UBR service classes). The use of VPCs may be used to segregate this traffic from traffic which is not tolerant of statistical multiplexing (such as CBR traffic).
- For connection endpoints served by VPCs, UPC/NPC functions may be performed on the aggregate traffic carried over the VPC.

8. Traffic Shaping

Traffic shaping is a mechanism that alters the traffic characteristics of a stream of cells on a VCC or a VPC to achieve desired modification of those traffic characteristics in order to achieve better network efficiency while meeting QoS objectives. Alternatively, traffic shaping may be used at the B-ICI to ensure conformance at a subsequent network interface. Shaping modifies traffic characteristics of a cell flow with the consequence of increasing the mean cell transfer delay.

Examples of traffic shaping are peak cell rate reduction, burst length limiting, and reduction of CDV by suitably spacing cells in time.

It is the choice of a network operator as to whether traffic shaping will be performed. In the absence of traffic shaping, it is the responsibility of the traffic source to the network to ensure that the transmitted cell traffic is compliant with the traffic contract at the risk of cell loss for any noncompliant traffic.

9. Selective Cell Discard

A congested network element may selectively discard cells which meet either or both of the following conditions:

1. Cells that belong to a noncompliant ATM connection.
2. Cell that have CLP = 1.

The objective is to protect the CLP = 0 traffic flow to the extent possible when congestion conditions exist within the network.

10. Frame Discard

If the network is forced to discard cells that are a part of an AAL-PDU, or frame, it is more efficient to discard all remaining cells associated with the frame since a single dropped cell will result in the entire frame being retransmitted. This determination must be made at the ATM layer; however, only AAL Type 5 allows the ATM layer to delineate frame boundaries. This is accomplished by inspecting the ATM User-to-User indication subfield within the PTI field of the ATM cell header.

When it can be employed, frame discard (also referred to as Partial Packet Discard) helps reduce network congestion. If a network does support frame discard, the user must request this treatment during connection establishment.

11. Guaranteed Frame Rate Service

Guaranteed Frame Rate (GFR) Service is an ATM Forum service definition that is intended for users who are unable to specify the traffic parameters for the ATM services discussed in this chapter [4]. In particular, GFR supports users who send frame-based traffic by granting QoS to the frame rather than individual cells into which the frame is segmented for transport over an ATM network. GFR is based upon the following parameters:

- Minimum Cell Rate (MCR)
- Maximum Frame Size (MFS): expressed in cells
- Maximum Burst Size (MBS): which should be greater than or equal to the MFS

As BT is a function of MBS (which is in turn a function of MFS), it defines a cell burst containing at least as many cells as will be required to transport the frame over the ATM network. Thus, upon arrival of the first cell in the frame, a conformance test determines whether the entire frame will be conforming or nonconforming. In the Leaky Bucket Algorithm the conformance test is:

$$X' > BT(MFS) + CDVT$$

A frame is considered nonconforming if any of its cells are considered nonconforming.

Another aspect of conformance has to do with the value of the CLP bit in the ATM cell header. For a user-generated cell to be conforming within the GFR service, all received cells must have the same CLP bit value as the first cell in the frame. The network may also (optionally) tag untagged user cells.

The network may discard all cells of any nonconforming frame although the intention of the GFR service is that nonconforming traffic be delivered if there are available resources within the network (thus, the inclusion of an MCR in the service definition). Cell discard, if necessary, in GFR service generally follows the procedure described for Frame Discard: if the first cell in the frame is the first cell discarded, then all remaining cells in the frame are to be discarded; otherwise all remaining cells after the first discarded cell are to be discarded, with the exception of the last cell, which provided frame delineation for the succeeding frame.

A GCRA for the GFR service may be defined as follows:

$$GCRA(T_{PCR, 0+1}, CDVT)$$

$$GCRA(T_{MCR, 0}, BT + CDVT)$$

References

1. Traffic Management Specification, Version 4.0, ATM Forum, Document af-tm-0056.000, April 1996.
2. B-ISDN ATM Layer Cell Transfer Performance, International Telecommunication Union, Recommendation I.356, October 1996.
3. Asynchronous Transfer Mode Management of the Network Element View, International Telecommunication Union, Recommendation I.751, March 1996.
4. Traffic Management Baseline Text Document, ATM Forum, Document BTD-TM-01.02, July 1998.

PART 3

CHAPTER 7

ATM Service Interworking

One of the roles for ATM is in providing wide area connectivity over public networks. Acceptance of ATM by users, however, requires that it be compatible with the embedded base of wide area services. Such interworking offers benefits to both users and public carriers. For public carriers, it allows them to make better use of their capital investment in ATM equipment by carrying more traffic than they would by serving only ATM customers. From the user perspective, interworking allows them to maintain their existing network interfaces while migrating to ATM interfaces over time. Ad interim, the customer is able to maintain full connectivity among sites whether those sites use ATM interfaces or not.

In this chapter we will discuss ATM interworking with three different types of existing public carrier services: Circuit Switched Services, Frame Relay Services, and Switched Multi-megabit Services. We will also describe an important variation of the ATM UNI known as the Frame Based UNI.

1. Background

There is an established demand for circuit-switched services for real-time, interactive types of traffic. Once the circuit-switched connection is established, the bandwidth associated with the end-to-end transmission path is dedicated to that connection, whether the channel is carrying active traffic or is idle. This is a desirable characteristic for interactive traffic such as voice and videoconferencing where one wants to avoid "broken" speech and visually unappealing "jagged" motion. Data traffic, which is more bursty in nature and which typically does not have real time requirements, is more tolerant of traditional packet-switched services where virtual connections are established and channel bandwidth used only when needed to carry active traffic.

Services for data communications have seen considerable evolution since the 1970s when X.25 services were first offered. X.25 provided packet-switched services for data rates between 2.4 kbps up to 64 kbps. At the time, the primary user need was terminal to host connectivity. Due to the relatively poor transmission facilities that existed at the time, bit error rates were such that extensive error checking was performed at each network node as the packet traffic travelled across an end-to-end connection. In the X.25 protocol there are approximately 30 error checking or other processing steps that must occur at every node [1]. This contributed to long packet transit delays that, in turn, limited the bit rates that could be supported by X.25.

By the mid 1980s, the need to connect widely dispersed local area networks produced increased traffic and a need for higher data rates. Users began to migrate traffic away from public networks and on to private networks using leased facilities. Routers at each user site would be connected to a Channel Service Unit/Data Service Unit (CSU/DSU) which provided network termination for the leased circuit on the customer premises. Private facilities, which offered data rates from 56 kbps to 1.544 Mbps, provided users with guaranteed bandwidth which ensured information managers that whenever they wanted send traffic from one site to another, they would have the bandwidth available to do so. This resulted in better throughput and response times for users. Private networks provided users with a certain measure of flexibility in deploying bandwidth; for instance, if a single DS-1 facility did not provide sufficient bandwidth, a second DS-1 private circuit could be leased between the two sites, providing an aggregate bit rate of 3.088 Mbps.

During this period, Integrated Service Data Network (ISDN) emerged as a new public network service [2]. ISDN supported bit rates from DS-0 to DS-1. However, delays in the establishment of national ISDN standards among Network Equipment manufacturers, and high costs for customer premises equipment and services, delayed the widespread acceptance of ISDN.

One advantage of a public network service over a leased line service is that public network services offer "one to many" connectivity; virtual connections may be established from one site to any of several sites over a single transmission facility from the customer premises. Private network services are "point to point;" a separate facility must be deployed to connect each pair of sites. For a large, fully connected network, the number of connections (and associated costs) may become prohibitive: for N sites, N * (N - 1)/2 facilities would be required to fully connect the network using leased facilities. Alternative topologies using fewer facilities are possible by using "hubbing." In hubbing, certain sites are designated as being hub sites. Hub sites form a fully connected mesh using private circuits. Nonhub, or "satellite" sites are then connected to the nearest hub. Traffic between hub sites travels in a single hop, while traffic between nonhub sites has at least one intermediate hop through the nearest hub.

In the early 1990s, developments in the public network arena, such as the deployment of optical fiber, resulted in much more reliable, lower bit error rate, transmission facilities that obviated the extensive error checking performed by X.25. Frame Relay Service emerged based upon the premise that, with bit errors so unlikely within the network, much less rigorous error checking was required. In Frame Relay networks, each node checks the integrity of the information in the Frame Header only. Unlike X.25, where error processing begins only after the entire packet has arrived at the node, Frame Relay can process the frame as soon as the frame header has arrived. This results in reduced transit delay, and, in turn, higher bit rates. Initial marketing of Frame Relay services by public carriers was targeted for data rates between DS-0, and DS-1 [3].

Shortly after the introduction of Frame Relay came Switched Multi-megabit Data Service (SMDS) which was targeted for LAN connectivity. SMDS supported data rates between DS-1 and DS-3. Unlike Frame Relay, which transports data in variable length frames, SMDS transports data in fixed length 53-octet cells. While this is the same size as ATM cells, there are two key differences:

1. SMDS is Connectionless. Unlike connection-oriented services, which travel over a predefined end-to-end route, connectionless services are not so constrained. Consequently, there could be variation in end-to-end transit delay due to individual cells taking different routes through the network.
2. SMDS has no Class or Quality of Service Specification. Even though there is a QoS field in the SMDS header, at this writing there is no definition of how this field is to be used. Thus, no distinction is made between different types of traffic in SMDS.

ATM provides support for private line, Frame Relay, and SMDS services through the use of Interworking Functions (IWFs), which carry traffic that originates or terminates on non-ATM networks and interfaces [4]. The switches that implement IWF capabilities are referred to as *edge devices*.

2. Circuit Emulation Service

ATM is a packet-oriented transmission technology. However, Constant Bit Rate (CBR) traffic with strict end-to-end timing requirements, is better carried using circuit-switched transmission technology. To provide service to the large installed base of circuit-switched transmission equipment, ATM must provide a means to emulate the characteristics of circuit-oriented transmission media.

Circuit Emulation Service (CES) provides ATM with the capability to provide performance comparable to that of Time Division Multiplexing (TDM) technology. CES allows subscribers to connect their legacy TDM equipment to ATM networks.

There are two types of CES [5]:

- Structured DS1/E1/J2, which provides point-to-point NxDS0 (64 Kbps), or "fractional" DS1/E1/J2 CES over ATM networks; and,
- Unstructured DS1/E1/J2/DS3/E3, which provides point-to-point 1.544 Mbps/2.048 Mbps/6.312 Mbps/44.736 Mbps/34.368 Mbps CES over ATM networks.

End-to-end CES is shown in Figure 7-1. End-user (CBR) equipment interfaces with an ATM CES Interworking Function (IWF) over a CBR service interface. The physical layer interface is either DSX-1 (for DS1), DSX-3 (for DS3), G.703 (for E1 or E3), or JT-G.703a (for J2).

The interworking function is responsible for interfacing TDM traffic from the CBR service interface to the ATM access interface. As shown in Figure 7-2, the IWF performs the following functions:

- Physical Layer interface with both TDM and ATM networks.
- Mapping: assigning bits (unstructured CES) or time slots (structured CES) from incoming traffic over the CBR service interface to locations within the ATM cell in coordination with the AAL1 Segmentation and Reassembly (SAR) function.
- ATM Adaptation Layer (AAL). CES uses AAL 1 [6].
- ATM Layers. Because of the nature of most constant bit rate traffic, loss of any cell must be avoided. All cells in CES are marked with high priority for cell delivery (CLP0).

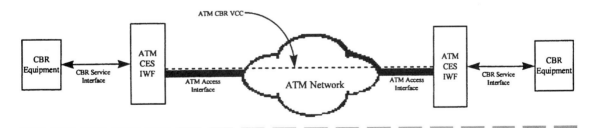

Figure 7-1. End-to-end ATM circuit emulation service.

Figure 7-2.
Circuit emulation service interworking function interface.

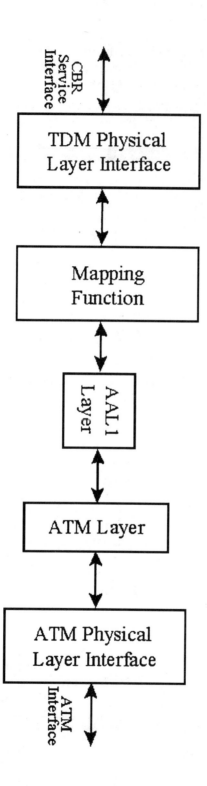

2.1 Structured CES

There are two types of Structured CES:

- Basic CES: in basic CES, time slots are handled transparently by the IWF and the ATM network. Basic CES provides no support for signalling. This type of Structured CES may be used where Common Channel Signalling is not required, or where signalling is provided by other means. In basic CES, ATM connections are PVC.
- CES with Channel Associated Signalling (CAS): in this type of Structured CES the end user equipment uses "robbed bit" signalling which is detected by the IWF. Signalling bits may also be detected but passed to the ATM network without interpretation by the IWF.

Figure 7-3 shows a Basic CES configuration. In this configuration, signalling and call control are provided by a Tandem switch external to the ATM network. PVCs route traffic between the IWF serving the Tandem switch and each of the other IWFs. Signalling is exchanged between the Tandem switch and end-user equipment connected to the CBR service interfaces transparent to the ATM network. Since the IWF performs no signalling function, each end-user is associated with fixed, permanent time slots in the DS1 Frame. Thus, when the Tandem switch receives signalling, it determines which time slots to connect at the originating and terminating IWFs to establish an end user to end-user connection.

One drawback to this approach is that for a large network, multiple Tandem switches would be employed. In a fully meshed network, each Tandem switch would be connected to every other Tandem switch. For an N Tandem switch network, this would require $N*(N-1)/2$ PVCs. Given that each ATM switch within the ATM network would have to be provisioned with this number of PVCs, this could result in a prodigious number of PVCs for large values of N.

Channel Associated Signalling With CAS, the CBR equipment at each CBR service interface may be connected without an external Tandem switch. In this context, signalling functionality within the IWF takes one of two forms:

1. Narrowband Signalling only. The originating IWF receives robbed bit signalling from a calling party over the CBR service interface and selects the IWF, which serves the called party. IWFs are connected through the ATM network using PVCs in a fully

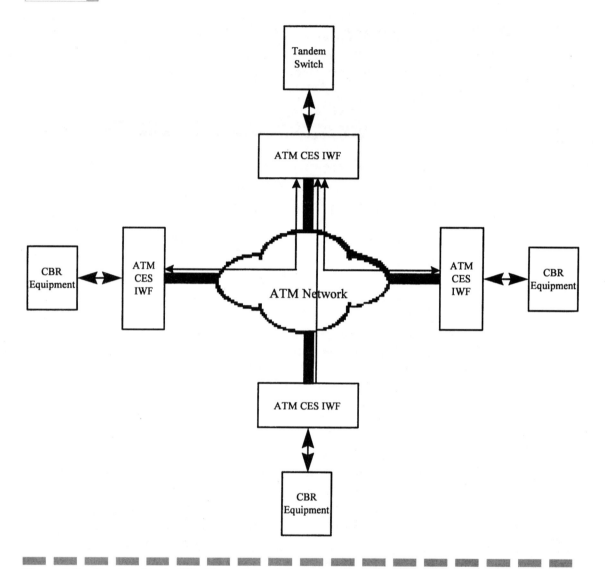

Figure 7-3. Structured basic CES with tandem switch.

connected mesh. Signalling messages are carried between the IWFs transparently through the ATM network using Channel Associated Signalling. The terminating IWF signals the called party over the CBR interface to establish the call.

2. Narrowband and Broadband Signalling. When the IWF receives signalling from the calling party it exchanges signalling within the ATM network to establish an SVC connection to the far end IWF (Broadband Signalling). Once this connection is established, the

IWFs exchange signalling across the CBR service interface (Narrowband Signalling, as described above) to establish a connection between the calling party and called party [9] [10].

Clocking For Structured CES the IWF provides a timing source to end user equipment across the CBR service interface. This serves as the primary clock from the network [11]. End-user equipment attached to this interface derives local clock timing from this IWF-supplied timing source (referred to as *loop timing*). In general, the IWF must have a high accuracy clock that is traceable to a Primary Reference Source (PRS).

2.1.1 Basic CES

In Basic CES the Mapping function groups timeslots received over the CBR service interface into AAL1 Service Data Units (SDU) containing N octets for NxDS0 rate subscriber interfaces. Figure 7-4 shows the mapping of a DS1 frame containing a 384 Kbps videoconference (N = 6), and a 64Kbps voice call. While time slots mapped into the AAL1 SDUs are shown to be contiguous in Figure 7-4, this is not necessary for the Mapping Function to distinguish one call from another. At the time of connection establishment, the ATM Management Plane forms an association between the ATM connection and specific time slots within the DS1 Frame.

As AAL1 SDUs are packed into ATM cells, delay is introduced due to the time it takes for enough AAL1 SDUs to arrive to fill an ATM cell. Since DS1 frames arrive every 125µs, this delay is approximately 6ms/N for NxDS0 connection. For a 64 Kbps voice, N = 1 and circuit cell assembly delay is approximately 6ms. One approach to reducing cell assembly delay is to partially fill each cell with user data with the rest of the cell being comprised of "dummy fill" octets.

Figure 7-5 gives an example showing how DS0 time slots for the 384 Kbps videoconference might map into ATM cells using Basic Structured CES. Note that there is only one P format cell out of every eight cells sent. P format cells are only sent when the AAL1 cell sequence number is even. For N = 1 there is no need for a structure pointer, thus all cells are non-P format.

2.1.2 CES with CAS

In CES with CAS, the Mapping function groups time slots into an AAL1 SDU referred to as a Payload Structure. This structure contains

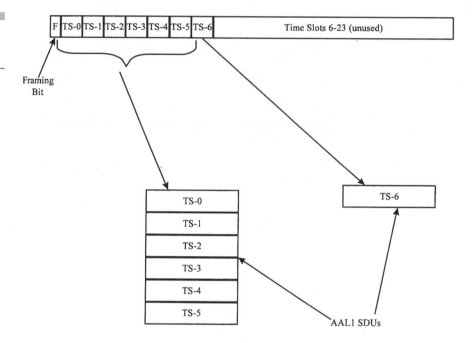

Figure 7-4.
Basic structured CES time-slot mapping.

time slots from 24 consecutive DS1 frames. Time slots are accumulated over 24 frames to accommodate the Extended Superframe (ESF) signalling format. During this time, 4 signalling bits are received for each time slot.

Within the Payload Structure shown in Figure 7-6, there is a Payload Substructure that contains 24 copies of the AAL1 SDU that would be produced in Basic CES. Thus, the Payload Substructure contains 24xN octets. In addition, the Payload Structure contains a Signalling Substructure that holds the signalling bits for each timeslot in the Payload Substructure. Each octet in the Signalling Substructure contains signalling bits from a pair of timeslots resulting in a Signalling Substructure that contains N/2 (rounded up to the next highest integer) octets.

Since 24 DS1 frames must arrive before an AAL1 SDU can be completed, this imposes a minimum delay of approximately 3ms.

Revisiting the videoconference example in Figure 7-7, the structure of the pointer that had a value of 4 in Figure 7-5d would now be 67, indicating that the beginning of the next structure block actually occurs in the cell following the one in which the structure pointer occurs.

Chapter 7: ATM Service Interworking

HDR	0	TS-0	TS-1	TS-2	TS-3	TS-4	TS-5	TS-0	TS-1	TS-2	TS-3	TS-4	TS-5	TS-0	TS-1
TS-2	TS-3	TS-4	TS-5	TS-0	TS-1	TS-2	TS-3	TS-4	TS-5	TS-0	TS-1	TS-2	TS-3	TS-4	TS-5
TS-0	TS-1	TS-2	TS-3	TS-4	TS-5	TS-0	TS-1	TS-2	TS-3	TS-4	TS-5	TS-0	TS-1	TS-2	TS-3

(Structure Pointer above "0"; Beginning of Structure Block above first TS-0)

a) First P format cell, beginning of "cycle of eight" with sequence count=0. Structure pointer indicates that the beginning of the next structure block is the octet following the SP. Within the SN field of the header octet, CPI=1 indicating that this is a P format cell.

HDR	TS-4	TS-5	TS-0	TS-1	TS-2	TS-3	TS-4	TS-5	TS-0	TS-1	TS-2	TS-3	TS-4	TS-5	TS-0
TS-1	TS-2	TS-3	TS-4	TS-5	TS-0	TS-1	TS-2	TS-3	TS-4	TS-5	TS-0	TS-1	TS-2	TS-3	TS-4
TS-5	TS-0	TS-1	TS-2	TS-3	TS-4	TS-5	TS-0	TS-1	TS-2	TS-3	TS-4	TS-5	TS-0	TS-1	TS-2

b) Sequence count=1, first non-P format cell. This cell contains no SP. The CPI bit is set to value 0.

HDR	TS-3	TS-4	TS-5	TS-0	TS-1	TS-2	TS-3	TS-4	TS-5	TS-0	TS-1	TS-2	TS-3	TS-4	TS-5
TS-0	TS-1	TS-2	TS-3	TS-4	TS-5	TS-0	TS-1	TS-2	TS-3	TS-4	TS-5	TS-0	TS-1	TS-2	TS-3
TS-4	TS-5	TS-0	TS-1	TS-2	TS-3	TS-4	TS-5	TS-0	TS-1	TS-2	TS-3	TS-4	TS-5	TS-0	TS-1

c) Sequence count=2, non-P format cell. This time CPI=0.

HDR	4	TS-3	TS-4	TS-5	TS-0	TS-1	TS-2	TS-3	TS-4	TS-5	TS-0	TS-1	TS-2	TS-3	TS-4
TS-5	TS-0	TS-1	TS-2	TS-3	TS-4	TS-5	TS-0	TS-1	TS-2	TS-3	TS-4	TS-5	TS-0	TS-1	TS-2
TS-3	TS-4	TS-5	TS-0	TS-1	TS-2	TS-3	TS-4	TS-5	TS-0	TS-1	TS-2	TS-3	TS-4	TS-5	TS-0

(Structure Pointer above "4"; Beginning of Structure Block above the TS-0 four octets after the SP)

d) Beginning of next "cycle of eight" with sequence count=0. This time the structure pointer indicates that the beginning of the next structure block is the fourth octet following the SP. Within the SN field of the header octet, CPI=1 indicating that this is a P format cell.

Figure 7-5. Structured data transfer example: Basic CES (ATM cell headers not shown). N = 6.

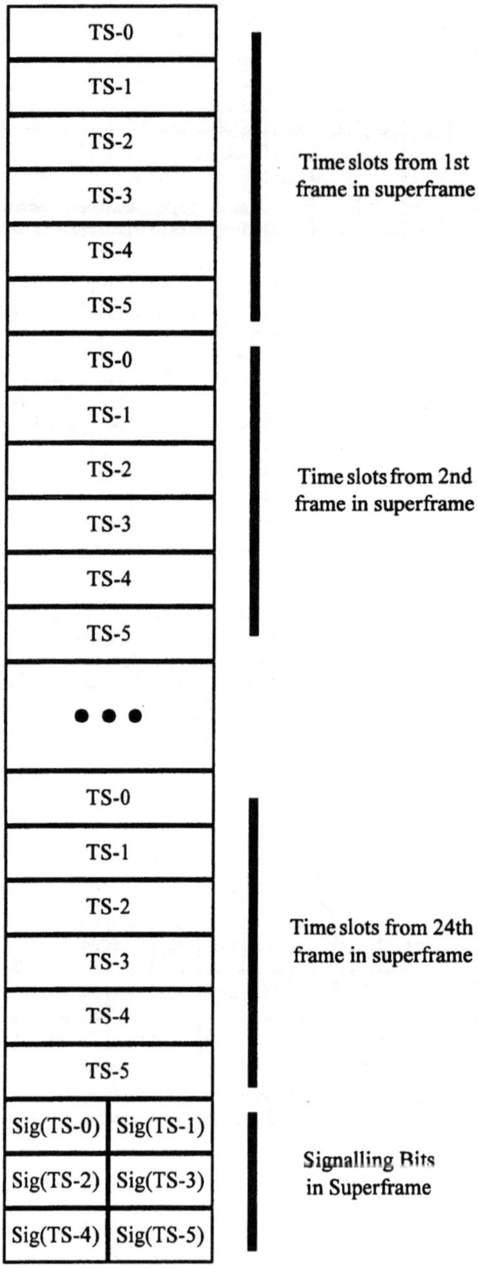

Figure 7-6.
CES with CAS: Time slot mapping into payload structure with N=6. The overall structure contains 147 octets.

Payload	Description
TS-0, TS-1, TS-2, TS-3, TS-4, TS-5	Time slots from 1st frame in superframe
TS-0, TS-1, TS-2, TS-3, TS-4, TS-5	Time slots from 2nd frame in superframe
...	
TS-0, TS-1, TS-2, TS-3, TS-4, TS-5	Time slots from 24th frame in superframe
Sig(TS-0) / Sig(TS-1); Sig(TS-2) / Sig(TS-3); Sig(TS-4) / Sig(TS-5)	Signalling Bits in Superframe

Chapter 7: ATM Service Interworking

	Structure Pointer	Beginning of Structure Block														
HDR	0	TS-0	TS-1	TS-2	TS-3	TS-4	TS-5	TS-0	TS-1	TS-2	TS-3	TS-4	TS-5	TS-0	TS-1
TS-2	TS-3	TS-4	TS-5	TS-0	TS-1	TS-2	TS-3	TS-4	TS-5	TS-0	TS-1	TS-2	TS-3	TS-4	TS-5
TS-0	TS-1	TS-2	TS-3	TS-4	TS-5	TS-0	TS-1	TS-2	TS-3	TS-4	TS-5	TS-0	TS-1	TS-2	TS-3

a) First P format cell, beginning of "cycle of eight" with sequence count=0.

	Structure Pointer														
HDR	67	TS-3	TS-4	TS-5	TS-0	TS-1	TS-2	TS-3	TS-4	TS-5	TS-0	TS-1	TS-2	TS-3	TS-4
TS-5	TS-0	TS-1	TS-2	TS-3	TS-4	TS-5	TS-0	TS-1	TS-2	TS-3	TS-4	TS-5	TS-0	TS-1	TS-2
TS-3	TS-4	TS-5	TS-0	TS-1	TS-2	TS-3	TS-4	TS-5	TS-0	TS-1	TS-2	TS-3	TS-4	TS-5	TS-0

b) Beginning of next "cycle of eight" with sequence count=0. This time the structure pointer indicates that the beginning of the next structure block is the 67th octet following the SP (i.e., the next cell).

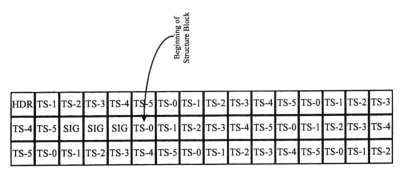

HDR	TS-1	TS-2	TS-3	TS-4	TS-5	TS-0	TS-1	TS-2	TS-3	TS-4	TS-5	TS-0	TS-1	TS-2	TS-3
TS-4	TS-5	SIG	SIG	SIG	TS-0	TS-1	TS-2	TS-3	TS-4	TS-5	TS-0	TS-1	TS-2	TS-3	TS-4
TS-5	TS-0	TS-1	TS-2	TS-3	TS-4	TS-5	TS-0	TS-1	TS-2	TS-3	TS-4	TS-5	TS-0	TS-1	TS-2

c) Sequence count=1, non-P format cell. The beginning of the next structure block begins in this cell as was indicated in the previous cell.

Figure 7-7. Structured data transfer example: CES with CAS (ATM cell headers not shown). N = 6.

2.1.3 ISDN CES

ISDN CES does not carry signalling information in the Payload Structure as is done with CAS, but rather uses a separate channel for signalling. Therefore, the AAL SDU is the same as that of Basic CES.

2.2 Unstructured CES

With relatively few configuration options, Unstructured CES is a simpler service to configure than Structured CES. Unstructured CES performs bit-by-bit mapping from the DS1 Frame to ATM cells. In this case, the AAL1 SDU is 1 bit. Figure 7-8 shows the mapping of DS1 frames into an ATM cell.

Clocking Unstructured CES has two timing modes for user equipment attached to the CBR Service Interface:

- Synchronous Mode: this is the same clocking arrangement as is used in Structured CES; master clocking is supplied by the IWF with user equipment loop timed to the IWF.
- Asynchronous Mode: in this mode, timing information between user equipment at each end of the connection is carried end-to-end through the ATM network. This method is called the Synchronous Residual Time Stamp (SRTS) Method.

Figure 7-8. Unstructured CES bit mapping.

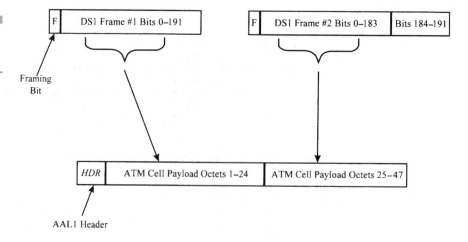

Chapter 7: ATM Service Interworking

SRTS allows the local clock at the user equipment to be (within limits) out of synchronization with the ATM network clock. The residual timestamp is a means for allowing the receiving end IWF to determine how out of synchronization (with respect to the Network clock) the sending user equipment is. On the sending end, the IWF compares the clock received over the CBR Service Interface with the Network clock, from which is derived a measure of the difference in the number of clock pulses observed between the two clocks—the residual time stamp. This time stamp, a 4 bit value (which allows approximately a +/− 8 clock pulse skew, or about +/− 5μs at DS1 clock rates), is encoded in the CPI bit in ATM cells when the sequence number is odd (MSB is sent on SN = 1, LSB when SN = 7).

An alternative to SRTS is the Adaptive Clock Method. With adaptive clocking the local IWF uses the amount of information received over an interface as an indication of the source clock frequency. This determination, however, is made locally; no end-to-end timing information is carried over the network. For instance, the IWF may look at the fill level in its input buffer, increasing the local clock rate when the fill level exceeds an upper threshold value and decreasing it when the fill level falls below a lower threshold value.

2.3 Variable Bit Rate CES

This service is intended for packetized video and audio, which have end-to-end timing requirements, such as teleconferencing and multimedia applications. While AAL2 was initially proposed as a means of supporting ATM Service Class B, AAL2 was ultimately defined such that end-to-end timing was not carried. Broadband signalling, however, makes it possible to use AAL1 to carry variable bit rate data in what is called the Dynamic Structure Sizing Method [12]. Dynamic Structure Sizing specifies procedures that can be used to increase or decrease the structure size on an existing ATM connection carrying structured CES—in essence changing the bandwidth utilization on the connection. Changes in bandwidth utilization on a connection implies changes in the IWF mapping function which must add or delete time slots mapped into the AAL SDU, and consequently changes in the size of the AAL SDU.

2.4 ATM Trunking

2.4.1 Channel Associated Signalling

The functionality of the IWF with Channel Associated Signalling termination capabilities is shown in Figure 7-9. Incoming DS1, E1, or J1 trunks (with their associated signalling channels) terminate on the Narrowband switch function of the IWF. This narrowband switch arranges bearer channel DS0 time slots in such a manner that all DS0 time slots that are to follow a common path through the ATM network exit the narrowband switch on the same port. The outputs from the narrowband switching function are then presented to a TDM Physical Layer function.

The narrowband switching function also extracts the "robbed bit" (call supervision) signalling, and the DTMF (call routing) signalling

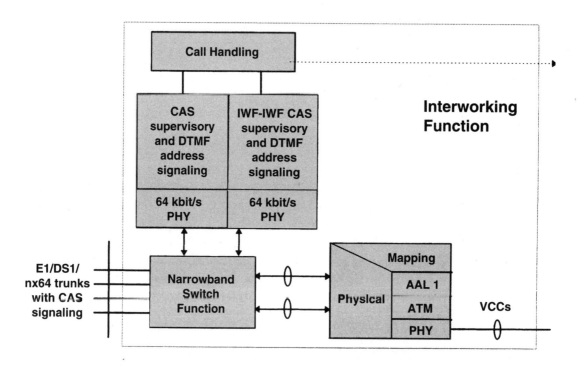

Figure 7-9. IWF functionality for support of channel associated signalling.

before forwarding them to a Call Processing function which effects signalling termination [20] [21]. The Call Processing function sends a message to the IWF-IWF signalling function which controls the narrowband switching function such that channel associated signalling is properly inserted into the outgoing (from the narrowband switching function) bearer traffic.

The narrowband switching function of the IWF allows the ATM SVC to carry a number of DS0s, possibly from multiple TDM connections, across the ATM network. This technique is referred to as *ATM Trunking*. The procedure by which DS0s from different connections are multiplexed into a single cell for transport across the ATM network is frequently referred to as Synchronous (Transfer Mode) to Asynchronous (Transfer Mode) Conversion, or SAC.

Figure 7-10 shows the information flows for the establishment of an ATM trunking SVC when no VCCs exist ab initio. The process of connection establishment is as follows:

1. The IWF receives an indication of Trunk Seizure (via robbed bit detection) from the narrowband network.

2. The IWF typically returns a "Wink" indication to indicate that the Calling Party may now begin dialing.

3. The Calling Party dials the digits to indicate the Called Party number, which is received by the originating IWF via DTMF signalling.

4. Based upon the "Called Party" number, the IWF maps that number to a destination address for the IWF which is to be the exit point from the ATM network for the call.

5. The IWF generates a Broadband ISDN Setup message, which is sent across the ATM network to the destination IWF over the well-known ATM signalling VCC. The objective of this Setup message is to establish the bearer path (which will also carry signalling information).

6. The receiving IWF accepts the narrowband signalling SVC setup request and returns a Broadband Connect message to the sending IWF.

7. Next the originating IWF forwards (via CAS) the Trunk Seizure indication to the receiving IWF.

8. The receiving IWF returns a Wink to the originating IWF.

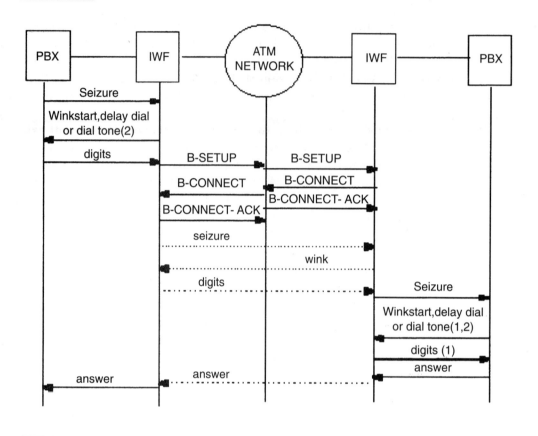

Figure 7-10. Information flows for channel associated signalling.

9. The originating IWF forwards the digits of the Called Party number to the receiving IWF.
10. Based upon the Called Party number, the receiving IWF seizes an outgoing trunk to the call terminating narrowband network. This is followed by a Wink and transmission of DTMF digits.
11. Assuming the Called Party answers, the receiving IWF receives indication of the event (also known as answer supervision) via CAS from the call terminating narrowband network.
12. The receiving IWF forwards answer supervision to the originating IWF.
13. The originating IWF forwards answer supervision to the call-originating narrowband network, which establishes the call.

2.4.2 Narrowband ISDN Signalling

The Q931 signalling procedures of ISDN are similar to the Q2931 signalling procedures of ATM signalling [7] [8]. There is, however, message translation that must be performed by the IWF to account for differences between the two networks. For instance, in ISDN signalling there are no provisions for QoS, or AAL specification so the IWF must insert or remove attributes from signalling messages when transferring signalling end-to-end between the ISDN and ATM signalling realms.

The functionality of the IWF with Narrowband ISDN signalling termination capabilities is shown in Figure 7-11. Incoming DS1 or E1 trunks (with their associated signalling channels) terminate on the Narrowband switch function of the IWF.

The narrowband switching function also extracts the Narrowband ISDN signalling channel and directs it to other IWF functions which extract the narrowband ISDN signalling messages before forwarding them to a Call Processing function which effects signalling termination. The Call Processing function sends a message to the IWF-IWF signalling function, which constructs the appropriate broadband ISDN signalling message, which is transported across the ATM network using the SAAL.

Figure 7-12 shows the information flows for the establishment of an ATM trunking SVC when no VCCs exist ab initio. The process of connection establishment is as follows:

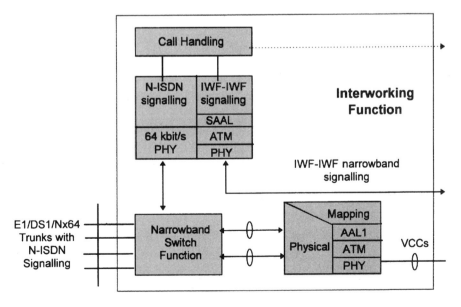

Figure 7-11.
IWF functionality for support of narrowband ISDN signalling.

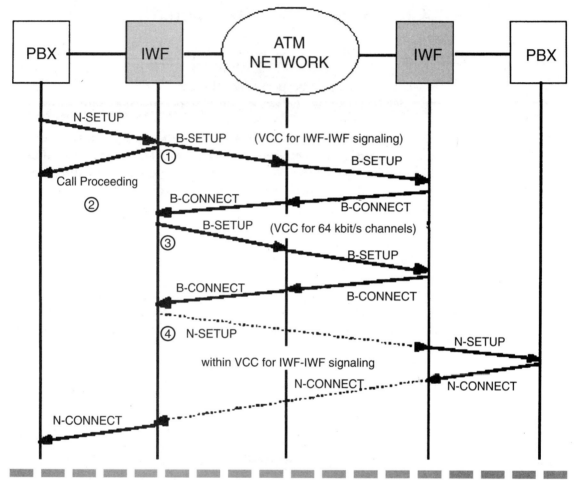

Figure 7-12. Information flows for narrowband ISDN signalling.

1. The IWF receives a Narrowband ISDN Setup message.
2. The IWF typically returns a Narrowband ISDN Call Proceeding message to prevent a time-out at the narrowband source while establishing a path through the ATM network.
3. Based upon the "Called Party" number in the Narrowband ISDN Setup message, the IWF maps that destination to a destination address for the IWF which is to be the exist point from the ATM network for the call.
4. The IWF generates a Broadband ISDN Setup message, which is sent across the ATM network to the destination IWF over the well-

known ATM signalling VCC. The objective of this Setup message is to establish a signalling path between IWFs for the exchange of Narrowband ISDN Signalling messages.

5. The receiving IWF accepts the narrowband signalling SVC Setup request and returns a Broadband Connect message to the sending IWF.

6. Next the originating IWF establishes the bearer path for the ATM trunk by sending another Broadband Setup message over the well-known ATM signalling VCC. Associated with this narrowband bearer SVC is a Connection Identifier (CID) which is sent between IWFs in the Broadband High Layer information element of the Broadband Setup message.

7. The receiving IWF accepts the narrowband bearer channel SVC Setup request and returns a Broadband Connect message to the sending IWF.

8. The originating IWF now forwards the Narrowband ISDN Setup message (received in Step 1 above) over the Narrowband signalling SVC (established in Step 5). This narrowband ISDN Setup message uses the CID of Step 6 in the Connection Identifier field.

9. When the receiving IWF receives the Narrowband ISDN Setup message, it relays the message over the Narrowband ISDN network to the Called Party.

10. Assuming the Called Party accepts the call, a Narrowband Connect message is sent to the receiving IWF.

11. The receiving IWF forwards the Narrowband Connect message over the Narrowband ISDN signalling SVC to the originating IWF.

12. The originating IWF, in turn, forwards the Narrowband Connect message over the Narrowband ISDN network to the Calling Party to establish the connection.

In the above example we see that there were separate VCCs for narrowband signalling traffic and narrowband bearer traffic. Since narrowband signalling is exchanged between IWFs transparent to the ATM network, the IWFs must have a mechanism for associating the exchange of messages over the narrowband signalling VCC, with the corresponding VCC, which is to carry the narrowband bearer traffic. This binding is accomplished by assigning a CID during narrowband bearer SVC Setup and subsequently using that CID in the Connection Identifier field of the narrowband ISDN Setup message. Note that this binding does not have significance outside of the ATM network.

2.5 CES Service Definition

In this section we relate items from the previous discussion to attributes that would be used to configure Structured and Unstructured CES.

Structured CES	
AAL Type	AAL1
CBR Rate	This is specified as being either: 64 Kbps, or Nx64 Kbps for N>1
Multiplier	The value for "N"; specified only for Nx64 Kbps service
Block Size	Specifies the size of the AAL1 SDU. As described earlier, the size of an AAL1 block depends upon the value "N", and whether Basic CES, or CES with CAS is used
Partial Cell Fill	If ATM cells are to be partially filled, this specifies the number of usable octets to insert in each ATM cell; specified only if a partial cell fill method is used
Protocol ID	Identifies the specific CES:
	— DS1/E1 Basic Service
	— E1 Service w/CAS
	— DS1 Service w/CAS and SF Framing
	— DS1 Service w/CAS and ESF Framing

Unstructured CES	
AAL Type	AAL1
CBR Rate	This is specified as being either: DS1 or E1
Clock Recovery Method	This specifies how clock is to be recovered by the IWF as being either asynchronous method, or one of two synchronous methods: SRTS, or Adaptive.

While not an exhaustive specification of a service contract, this does serve to give the reader a sense of the differences between how the two services would be specified.

The bit rate for the CES must be translated into a cell rate. In Structured CES, one of eight cells contains 46 octets of user data while the other seven cells contain 47 octets. This corresponds to an average cell size of 47.875 octets. The cell rate for Structured CES (without CAS) is then:

Cell Rate = (Aggregate CBR Rate)/(46.875 octets/cell * 8 bits/octet)

For Unstructured CES, where all cells contain 47 octets of user data, the cell rate is:

$$\text{Cell Rate} = (\text{Aggregate CBR Rate})/(47 \text{ octets/cell} * 8 \text{ bits/octet})$$

3. Frame Relay Service

Frame Relay Service (FRS) is a connection-oriented data transport service that provides for the transfer of variable length packets [13]. Typical applications are LAN interconnection over DS0 and DS1 interfaces. At the time of this writing, FRS is defined as a PVC-only service [14].

Frame Relay is an ISDN packet mode service. Frame relay traffic is interfaced to an ATM network across a Frame Relay Interworking Point (FR-IWP) at a Frame Relay-to-ATM IWF. The interworking function is responsible for interfacing Frame Relay traffic to the ATM access interface. The physical layer interfaces are ISDN Primary Rate Interface (PRI), and ISDN Basic Rate Interface (BRI) for DS1 and DS0 rates, respectively. As shown in Figure 7-13, the FRS IWF performs the following functions:

1. Physical Layer interface with both Frame Relay and ATM networks.

2. Frame Relay Service Specific Convergence Sublayer (FR-SSCS) provides mapping functions that interpret bits in the Frame Relay frame header and translates them to values in the ATM cell header (and vice versa) [4] [15].

3. AAL. FRS uses AAL5.

4. ATM Layer.

The FR-SSCS in the IWF can also multiplex traffic from more than one frame relay connection into a single ATM Virtual Channel Connection. These frame connections may be physically carried over a single FR-IWP or multiple FR-IWPs. Frame Relay connections are identified by the Data Link Connection Identifier (DLCI) field in the frame header.

3.1 Frame Structure

Figure 7-14 shows the default 2-octet header format of a Frame Relay frame (other formats allow 3 or 4 octet header fields). The fields in the default frame header are:

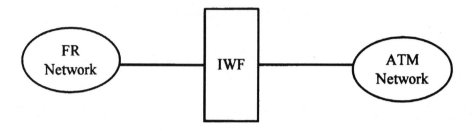

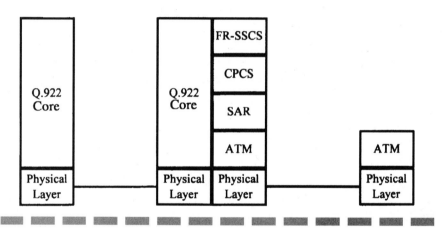

Figure 7-13. Frame Relay to ATM network interworking.

- DLCI (10 bits in Figure 7-14, but can be as many as 23 bits), which identifies the Frame Relay virtual channel.
- Command/Response (C/R) bit: is equal to value "0" if the frame contains a command, and "1" when the frame contains a response.
- Address Field Extension (EA) bit: indicates that there is another address octet following when set to "0" in the first (and any subsequent) octet. When set to "1," indicates that this is the last octet of the address field.
- Forward Explicit Congestion Notification (FECN) bit. When set to "1" in frame, it indicates that network congestion was experienced

in the "forward" (sender to receiver) direction. The intention of this bit was for user equipment receiving this frame to use it as an indication to decrease its data transfer rate until the congestion condition is cleared. In practice, however, most LAN attached equipment has no mechanism for using this information.

- Backward Explicit Congestion Notification (BECN) bit. This is a notification inserted in "backward" traffic to inform a sender that traffic being sent from the user equipment was encountering network congestion as it was being sent to the recipient. When set to "1" in frame, it indicates that traffic being sent from this node was experiencing network congestion in the (sender to receiver) direction. The intention of this bit was for user equipment that was the source of the traffic

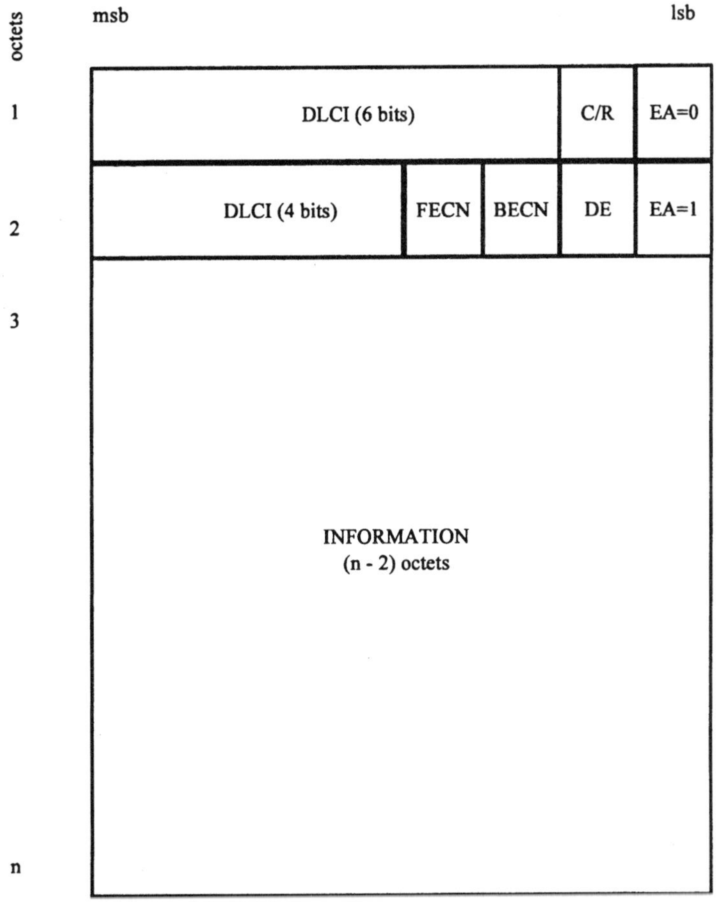

Figure 7-14.
Frame relay structure (default header format).

octets	msb				lsb
1	DLCI (6 bits)			C/R	EA=0
2	DLCI (4 bits)	FECN	BECN	DE	EA=1
3					
	INFORMATION (n - 2) octets				
n					

experiencing network congestion to use this as an indication to decrease its data transfer rate until the congestion condition is cleared. Most LAN attached equipment has no mechanisms for using this information either.

- Discard Eligibility (DE) indicator, which, when set to value "1," indicates that this frame is to be sent with low delivery priority. In the event that the network must discard frame traffic, frames marked with this value will be discarded before frames with DE value "0," which are sent with higher delivery priority.

3.2 Frame Relay Service Specific Convergence Sublayer

3.2.1 Discard Eligibility (Frame) to Cell Loss Priority (Cell) Mapping

Figure 7-15 shows how the Transmit Side IWF maps the DE bit from a frame to the CLP bit in a cell for traffic in the Frame Relay to ATM Network direction. At the originating IWF the value of the DE bit from the frame is copied to the CLP bit in every ATM cell used to transmit the frame through the ATM network. A network provider may, as an option, choose to ignore the DE bit and set the CLP bit to a constant value (either "0," or "1") for all cells.

Figure 7-15 also shows how the destination IWF maps CLP to DE for traffic in the ATM Network to Frame Relay Network direction. In this case, the DE of the outgoing frame is the logical OR of the DE bit of the received (from the Frame Relay network) frame and the CLP bits of each cell used to carry the frame through the ATM network. Note that the DE bit is carried through the ATM network in the first cell. Regardless of the value of the CLP bit as set by the originating IWF, this value may change (the only change would from high priority to low priority) as the cell transits the ATM Network. The network provider does, however, have the option of not mapping CLP bits to the DE bit in an outgoing frame in which case, the frame sent out from the ATM network has the same DE value as the frame received from the Frame Relay network. Here, the option would be to completely ignore (from the Frame Relay perspective) conditions that exist in the ATM network.

Chapter 7: ATM Service Interworking

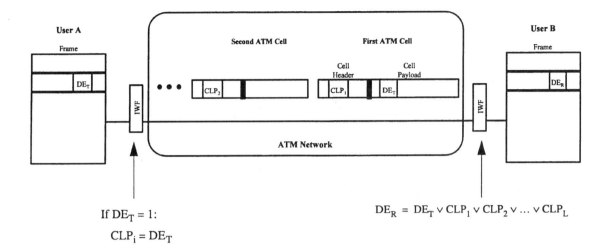

Figure 7-15. DE to CLP mapping.

3.2.2 Forward Explicit Congestion Notification (Frame) To Explicit Forward Congestion Indication (Cell) Mapping

EFCI is the ATM Cell-equivalent to FECN within a Frame. The EFCI bit is contained within the PTI field of the ATM cell header.

The IWF does not map the FECN bit to the EFCI bit for traffic in the Frame Relay to ATM Network direction; all cells enter the ATM network with EFCI=0. The value of the FECN bit from the incoming (to the ATM network) frame is always sent across the ATM network with the rest of the frame.

In Figure 7-16 we see the mapping of the EFCI bit to the FECN bit for traffic in the ATM Network to Frame Relay Network direction. In this context, only the EFCI bit value of the *last* ATM cell used to carry the frame through the ATM network is relevant. The reason for ignoring all other cells is because the last cell gives the most current

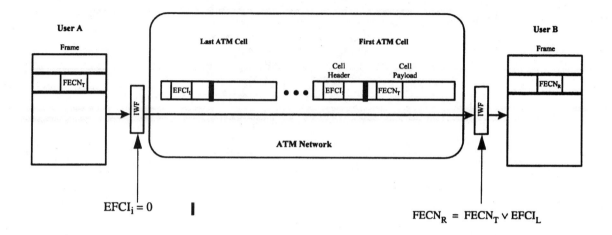

Figure 7-16. FECN to EFCI mapping.

FECN$_R$ - Forward Explicit Congestion Notification Bit in Received Frame
FECN$_T$ - Forward Explicit Congestion Notification Bit in Transmitted Frame
EFCI - EFCI Bit in i^{th} ATM Cell
EFCI$_l$ - EFCI Bit in last ATM Cell

information on the congestion state within the ATM network. If intermediate cells indicated congestion but the last cell indicated no congestion, then we may conclude that the congestion was a transient condition. As shown in Figure 7-16, the value of FECN in the outgoing frame is the logical OR of the EFCI bit in the last ATM cell and the FECN in the received frame.

3.2.3 Backward Explicit Congestion Notification (Frame) To Explicit Forward Congestion Indication (Cell) Mapping

There is no equivalent to the frame BECN in the ATM cell. Thus, in the ATM network to Frame Relay network direction, the BECN in the received frame is copied to the outgoing frame.

Figure 7-17 shows how the IWF maps the BECN bit from the Frame to the EFCI bit in the ATM cell for traffic in the Frame Relay to ATM network direction. If the BECN bit in the incoming (to the ATM network) frame indicates congestion, then this value is carried in the frame as it is sent through the ATM network (this bit is sent in the first ATM cell). The BECN bit is also set in the incoming frame as it is transported over the ATM network if the last cell of the last frame sent across the ATM network in the opposite direction had its EFCI bit set. This is the case of "backward" notification; a frame was previously sent in the opposite direction with forward explicit congestion notification, and the ATM network sends a backward indication the next time it has an opportunity (i.e., when it next sends traffic to the source of the FECN marked frame).

3.3 Frame Relay and ATM Interworking: Network View and Service View

3.3.1 Implementation Agreement FR.F5

Frame Relay/ATM service interworking may be viewed from two perspectives:

1. The ATM network provides interconnection between Frame Relay networks (Figure 7-18)
2. The ATM network interworks a Frame Relay service subscriber to an ATM service subscriber (Figure 7-19)

When the ATM network is used as a transit network for Frame Relay network interconnection, the IWF tunnels frames across the ATM network as described in the last section. The IWF also allows the Frame Relay header to be updated based upon traffic conditions within the ATM network. When used to interconnect a Frame Relay subscriber to an ATM subscriber, however, the IWF requires that the ATM subscriber terminal emulate a Frame Relay terminal in the FR-SSCS. The specification which defines an IWF which tunnels Frame Relay frames across ATM networks is commonly known as *Frame Relay/ATM Network Interworking Implementation Agreement FR.F5* [22].

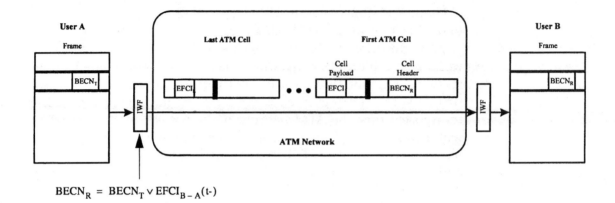

$BECN_R = BECN_T \lor EFCI_{B-A}(t-)$

$BECN_R$ - Backward Explicit Congestion Notification Bit in Received Frame
$BECN_T$ - Backward Explicit Congestion Notification Bit in Transmitted Frame
$EFCI_{B-A}(t-)$ - EFCI Bit of last ATM Cell sent from User B to User A

Figure 7-17. BECN to EFCI mapping.

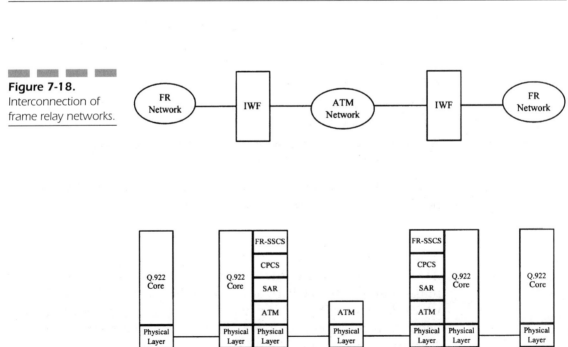

Figure 7-18. Interconnection of frame relay networks.

Figure 7-19.
Interworking of frame relay subscriber to ATM subscriber.

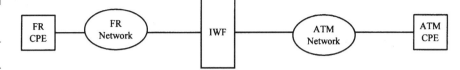

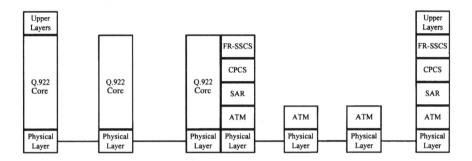

3.3.2 Implementation Agreement FRF.8

A second implementation agreement, *Frame Relay/ATM Service Interworking Implementation Agreement FRF.8*, defines an IWF, which acts as a protocol converter from frames to cells [23]. This allows interconnection of ATM subscribers and frame relay subscribers with neither being aware of the type of service being subscribed by the other. Thus, the ATM subscriber terminal is able to use a null SSCS.

For traffic entering the ATM network from a Frame Relay network, an FRF.8 IWF transcodes frame header information into ATM cell headers for transport across the ATM network just as does an FRF.5 IWF. There are two key differences between the transcoding methods of FRF.5 and FRF.8:

1. FRF. 8 strips the header of an incoming (from the Frame Relay network) frame. In the opposite direction, the FRF.8 IWF constructs a frame header based upon ATM cell header information.
2. FRF. 8 uses different rules in mapping between frame header bits and cell header bits
    a) DE to CLP mapping—same as FRF.5.
    b) FECN to EFCI mapping—FECN is mapped to the EFCI bit in every cell in the Frame Relay to ATM direction. In the opposite

direction the EFCI of the last cell is mapped to the FECN of the outgoing frame.

c) BECN is not mapped in FRF.8. BECN is ignored in incoming frames and set to value "0" in outgoing frames.

d) The Frame Relay C/R field is mapped to the CPCS-UU field.

e) The Frame Relay DLCI is mapped to an ATM VPI/VCI.

In addition to the above, an FRF.8 IWF performs Link Integrity Verification (LIV) to transfer information about the state of the links on either side of the IWF. On the Frame Relay network side, the IWF is to be able to poll the network and to respond to polls from the network [24]. If the IWF detects a service affecting condition in the ATM network (e.g. AIS/RDI cell detected) the IWF is to send a status report to the Frame Relay network. If a service affecting condition is detected in the Frame Relay network, an AIS F5 OAM cell is to be sent on all affected ATM PVCs.

3.4 FRS Service Definition

Frame Relay supports bursty traffic, thus, the subscriber must understand the traffic patterns that are to be carried over the network. A Frame Relay service definition is characterized by the following parameters [16]:

CIR   The Committed Information Rate, which represents the rate of sustainable traffic that can be sent across an FR-IWP. CIR equals Bc/T.

B_c The number of "credits," or bits that a user is allowed to send across the FR-IWP over a measurement interval, T. This parameter is normally not specified by the user but is set by the carrier.

B_e The number of excess credits (above Bc) that a user is allowed to send across the FR-IWP over a measurement interval, T. Like Bc, this parameter is normally specified by the carrier. Frame Relay allows users to send traffic at rates up to the access link rate. The Frame Relay service parameters limit the amount of traffic that can be transmitted at this rate, or any rate above the CIR. These parameters can be emulated within ATM FRS according to the application. Since LAN attached equipment does not respond to congestion information sent across the network it is the responsibility of the network operator to monitor the rate at which traffic is submitted to the network, taking corrective actions as necessary.

An example of a possible ATM FRS specification for a symmetrical, bidirectional connection is as follows:

Chapter 7: ATM Service Interworking

AAL Type	AAL5
SSCS Type	Frame Relay
Maximum CPCS SDU	The maximum size SDU for the connection; equal to the maximum frame size (which is determined when the frame relay connection is established). Specifications for the forward and backward directions are equal.
Peak Cell Rate (CLP = 0 + 1)	Aggregate peak cell transfer rate for CLP = 0 and CLP = 1 traffic. Approximately equal to the FR-IWP Access Link Rate (bps)/[48 (octets per cell) x 8 (bits/octet)]. For a DS1, PCR is 4,021 cells/second. Specified to be equal in both forward and backward directions.
Sustainable Cell Rate (CLP = 0)	Sustainable cell transfer rate for CLP = 0 traffic. Approximately equal to CIR (bps)/[48 (octets per cell) x 8 (bits/octet)].
Maximum Burst Size (CLP = 0)	Aggregate maximum number of CLP = 0 cells that can be transmitted in a burst at the PCR. Approximately equal to Bc (bps)/[48 (octets per cell) x 8 (bits/octet)]. Specified to be equal in both forward and backward directions.
Maximum Burst Size (CLP = 0 + 1)	Aggregate maximum number of CLP = 0 and CLP=1 cells that can be transmitted in a burst at the PCR. Approximately equal to [Bc + Be (bps)]/[48 (octets per cell) x 8 (bits/octet)]. Specified to be equal in both forward and backward directions. When taken with MBS (CLP = 0) specification, this specification allows Be "credits" of CLP=1 traffic to be sent at the PCR.
Cell Tagging	Allows nonconforming CLP = 0 cell traffic to be marked CLP = 1 (as opposed to being dropped). Conforming CLP = 0 cell traffic conforms to the PCR (CLP = 0 + 1) and SCR (CLP = 0) specifications.

4. ATM Frame Based User to Network Interface

Frame Relay is a more efficient protocol than ATM because for large payloads, Frame Relay overhead constitutes a smaller percentage of the overall transmission than is the case with ATM. In order to enable ATM users to realize greater protocol efficiencies across the UNI, the ATM Forum defined a Frame Based User to Network Interface (FUNI) Specification [25].

FUNI defines an interface by which frames may be passed across the FUNI to an ATM network as shown in Figure 7-20. FUNI specifies a FUNI to an ATM conversion function which converts FUNI frames into ATM cells. ATM terminals are able to connect to FUNI compatible terminals without a SSCS. At the time of this writing FUNI is defined for access rates of DS-1/E-1, and (optionally) higher rates.

FUNI frames are not the same as Frame Relay frames. For comparison the 2 octet and 4 octet FUNI frame header formats are shown in Figure 7-21. The following is the mapping between FUNI frames and ATM cells performed by the Conversion Function:

- Frame Address (FA)—in the 2 octet header bits 6-3 of the first FUNI header octet are mapped to the 4 LSBs of the VPI field in the ATM cell header. The 4 MSBs of the VPI field are set to 0. Bits 8-7 of octet 1, and bits 8-5 of octet 2 in the FUNI header are mapped to the 6 LSBs of the VCI field in the ATM cell header. The 10 MSBs of the VCI field are set to 0. In the 4-octet FUNI header bits 8-3 of the first header octet, and bits 8-7 of octet 2 are mapped to the VPI field of the ATM cell header. Bits 6-5 of the second FUNI header octet, bits 8-1 of the third octet, and bits 8-1 of the fourth octet are mapped to the VCI field of the ATM cell header.

- Frame Identification (FID1, FID2)—These bits are used to distinguish frames carrying user data from frames carrying OAM data. FID1 = FID2 = 0 for user data, and FID = 1, FID = 0 identifies an OAM frame. Other combinations are not defined.

- CLP—There is a direct mapping between FUNI frame header and ATM cell header.

- Congestion Notification (CN)—There is a direct mapping with the EFCI in the ATM cell header. In the network to user direction the CN is set equal to the value of the EFCI of the last constituent ATM cell. In the user to network direction CN is always equal to "0."

Figure 7-20.
FUNI reference model.

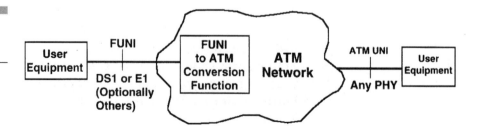

Figure 7-21.
FUNI frame header formats.

| Frame Address (6 bits) | FID1 | 0 |
| FA (4 bits) | CN | FID2 | CLP | 1 |

2 Octet Header

Frame Address (6 bits)	FID1	0
FA (4 bits)	CN	FID2	CLP	1
Frame Address (7 bits)	0
FA (7 bits)	1

4 Octet Header

FUNI supports AAL 5 (and optionally AAL 3/4) only. Furthermore, FUNI supports traffic management service classes nrt-VRT, and UBR only. As FUNI transports frames across an access link, GFR may also be used to ensure compliance of transported frames.

5. Switched Multi-megabit Data Service

Switched Multi-megabit Data Service (SMDS) is a connectionless data transport service that provides for the transfer of variable length packets (up to a maximum of 9,188 octets) [17]. SMDS is defined as a service in

Bellcore Technical Reference TR-TSV-000772 and, as such, it does not depend upon any particular communications implementation. In fact, the SMDS specification has been suggested as the method of providing Connectionless Broadband Data Services based upon ATM switching technology [18][19].

SMDS is targeted for LAN interconnection at DS1 and DS3 rates. SMDS access devices send data in 53-octet cells. While the same size as ATM cells, the format is different and is defined in the SMDS Interface Protocol (SIP). AAL 3/4 was defined to provide compatibility between ATM cells and SMDS cells.

The user interface for SMDS access is defined as the Subscriber Network Interface (SNI) and is based upon the IEEE 802.6 Dual Queue Dual Bus (DQDB) architecture.

The SMDS to ATM interworking function is responsible for interfacing SMDS traffic across an SNI to an ATM access interface. This IWF must interwork what is essentially a connectionless service (SMDS), with a connection-oriented network (ATM). The SMDS IWF performs the following functions:

1. Physical Layer interface between SMDS and ATM networks.
2. Encapsulation of the SIP Level 3 PDU (L3_PDU) within an Inter Carrier Interface Protocol Connectionless Service (ICIP CLS) PDU.
3. Mapping: responsible for mapping connectionless SMDS traffic to ATM connections.
4. Format Conversion: the SMDS IWF converts ICIP_CLS PDUs to AAL 3/4 CPCS PUs.
5. ATM Layer.

5.1 SIP L3_PDU Encapsulation

The SIP L3_PDU is comprised of the following fields:

- Reserved (Rsvd) (1 octet): populated with zeros; matching fields in the L3_PDU Header and Trailer.
- Beginning-End Tag (BEtag) (1 octet): marks the beginning and end of the L3_PDU. One instance of this field is placed in the Header and another in the Trailer, both having the same value.

- Buffer Allocation Size (BAsize) (2 octets): indicates the size (in octets) of the L3_PDU as measured between the BAsize field and the Trailer.
- Destination Address (DA) (8 octets). The address of the receiver of this L3_PDU.
- Source Address (DA) (8 octets). The address of the sender of this L3_PDU.
- Higher Layer Protocol Identifier (HLPI) (6 bits). Not processed by the network, exists solely to ensure alignment of the SIP format with the DQDB protocol format.
- Pad Length (PL) (2 bits): the length of the PAD field required to ensure that the L3_PDU is 32-bit aligned.
- Quality of Service (QoS) (4 bits). Not processed by the network, exists solely to ensure alignment of the SIP format with the DQDB protocol format.
- CRC32 Indication Bit (CIB). When set, indicates that a 4-octet CRC field is present in the L3_PDU.
- Header Extension Length (HEL) (3 bits): indicates the number of 32-bit segments in the Header Extension field. Set to value 3.
- Bridging (Brdg) (2 octets). Not processed by the network, exists solely to ensure alignment of the SIP format with the DQDB protocol format.
- Header Extension (HE) (12 octets): allows for extension of SMDS to accommodate additions or changes to SMDS features or access characteristics. Always 12 octets, this field may be (in whole or in part) PAD.
- Information (Info) (9,188 octets). User information field.
- PAD (3 octets). PAD field length is specified in PL field.
- Cyclical Redundancy Check (CRC32) (32 bits): performs error checking on L3_PDU fields between BAsize and CRC32. Presence in L3_PDU is indicated in the CIB field.
- Length (2 octets): has same value as BAsize field.

The SIP L3_PDU Encapsulation function truncates the L3_PDU trailer (Rsvd, BEtag, and Length fields at the end of the L3_PDU) to form the ICIP_CLS PDU.

5.2 SMDS to ATM Mapping

SMDS to ATM Mapping supports mapping between SMDS and ATM. These include:

- Address Mapping: associates a Source Address/Destination Address pair with a VPI/VCI.
- Group Addressing: is an SMDS feature that allows a logical (or "group") destination address to serve as a synonym for multiple physical destination addresses. This is equivalent to the LAN broadcasting. In this case, the Mapping function associates a Source Address/Group Address with multiple VPI/VCIs.

5.3 ATM Adaptation Layer

The Common Part Convergence Sublayer (CPCS) of AAL3/4 receives the ICIP_CLS PDU and forms a CPCS PDU by appending the following header fields:

- Common Part Indication (CPI) (1 octet): populated with zeros to indicate that the BAsize and Length fields are encoded in octets.
- Beginning Tag (Btag) (1 octet): marks the beginning of the CPCS PDU.
- Buffer Allocation Size (BAsize) (2 octets): indicates the size (in octets) of the CPCS PDU as measured between the BAsize field and the Trailer.

and the following trailer fields

- Alignment (AL) (1 octet): provides 32-bit alignment in the Trailer. Populated with zeros.
- End Tag (Etag) (1 octet): marks the beginning of the CPCS PDU. Value is the same as in Btag field.
- Length (2 octets): has the same value as BAsize field.

Within the CPCS PDU the ICIP_CLS PDU is considered as the Payload field. Overall, the CPCS PDU maintains 32-bit alignment. Since the ICIP_CLS PDU is already 32-bit aligned, and the CPCS Header and CPCS Trailer are 32-bit aligned, no PAD field is required in the Payload field.

The AAL3/4 SAR PDU is of the format of the SIP Level 2 PDU (minus the 5 octets of Access Control and Network Control Info in the SIP L2_PDU). In ATM, the MID field is used to associate a SAR PDU with a CPCS PDU. This effectively allows multiple logical connections on a single virtual connection within ATM. A similar relationship holds in SMDS where the MID field associates a SIP L2_PDU with a SIP L3_PDU. In the context of SMDS, the MID serves a role similiar to that of the Terminal Endpoint Identifier (TEI) in ISDN. In this way, SMDS can serve multipoint configurations in which multiple CPEs share the same access interface. The MID for the SIP L2_PDU need not be the same as that of the SAR PDU.

5.4 SMDS Service Definition

SMDS has no explicit flow control. Instead, a set of access classes is defined, each having a set of parameters. A "credit manager" algorithm enforces the rate of information flow based upon these parameters. An SMDS service definition is characterized by the following parameters [normally specified by the carrier based upon access class]:

C_{max} The maximum number of "credits," or octets that a user is allowed to accrue. This is generally set to value 9,188 for access classes in which the data rate is less than the maximum allowed on an access link.

I_{inc} The interval between increments to the credit. I_{inc} is measured by counting the number of SIP L2_PDUs (full and empty) received across the SNI.

N_{inc} The number of credits incremented every I_{inc} interval.

Note that while SMDS allows users to send traffic at rates up to the SNI access link rate, N_{inc} and I_{inc} define the Sustained Information Rate (SIR), which is the average rate at which user information may be transferred across the SNI. The SIR is the SMDS-equivalent to CIR in Frame Relay.

An example of a possible ATM SMDS specification for a symmetrical, bidirectional connection is as follows:

AAL Type	AAL3/4
SSCS Type	Null Encapsulation
Maximum CPCS SDU	The maximum size SDU for the connection; equal to the maximum ICIP_CLS PDU size (which is 9,231 octets). Specifications for the forward and backward directions are equal.

Peak Cell Rate (CLP = 0)	Peak cell transfer rate for CLP = 0 traffic. Approximately equal to the SNI Access Link Rate (bps)/[44 (octets per cell) x 8 (bits/octet)]. For a DS3 SNI, PCR is 127,841 cells/second. Specified to be equal in both forward and backward directions.
Sustainable Cell Rate (CLP = 0)	Sustainable cell transfer rate for CLP = 0 traffic. Approximately equal to SIR (bps)/[44 (octets per cell) x 8 (bits/octet)].
Maximum Burst Size (CLP = 0)	Aggregate maximum number of CLP = 0 cells that can be transmitted in a burst at the PCR. Approximately equal to Cmax (bps)/[44 (octets per cell) x 8 (bits/octet)]. Specified to be equal in both forward and backward directions.

References

1. Kessler, Gary, and David Train, *Metropolitan Area Networks: Concepts, Standards, and Services,* McGraw-Hill, Inc., 1992.

2. Kessler, Gary, and Peter Southwick, *ISDN: Concepts, Facilities, and Services,* 4th ed, McGraw-Hill, Inc., 1998.

3. Nguyen, Bach, "Frame Relay-Setting the Record Straight," *Business Communications Review,* October 1993.

4. BISDN Inter Carrier Interface (B-ICI) Specification, Version 2.0 (Integrated), ATM Forum, Document af-bici-0013.003, December 1995.

5. Circuit Emulation Service Interoperability Specification, Version 2.0, ATM Forum, Document af-vtoa-0078.000, January 1997.

6. B-ISDN ATM Adaptation Layer (AAL) Specification, Types 1 and 2, International Telecommunication Union, Document I.363, November 1994.

7. Digital Subscriber Signalling System No. 1 (DSS 1)—ISDN User-Network Interface Layer 3 Specification for Basic Call Control, International Telecommunication Union, Document Q931, March 1993.

8. Broadband Integrated Services Digital Network (B-ISDN)—Digital Subscriber Signalling System No. 2 (DSS 2)—User-Network Interface Layer 3 Specification for Point-to-Point Call/Connection Control, International Telecommunication Union, Document Q2931, February 1995.

9. Voice and Telephony Over ATM—ATM Trunking using AAL1 for Narrowband Services Version 1.0, ATM Forum, Document af-vtoa-0089.000, July 1997.

10. General Arrangements For Interworking Between BISDN and 64 kbit/s Based ISDN, International Telecommunications Union, Document I.580, March 1993.

11. Reeve, Witham, Subscriber Loop Signalling and Transmission Handbook, Digital, IEEE Press, 1995.

12. Specifications of (DBCES) Dynamic Bandwidth Utilization—In 64kbps Time Slot Trunking Over ATM Using CES, ATM Forum, Document af-vtoa-0085.000, July 1997.

13. ISDN Data Link Layer Specification for Frame Mode Bearer Services, International Telecommunication Union, Document Q922, 1992.

14. Frame Relaying Bearer Service Network-to-Network Interface Requirements, International Telecommunication Union, Recommendation I.372, March 1993.

15. Frame Relaying Service Specific Convergence Sublayer, International Telecommunication Union, Document I.365.1, November 1993.

16. Congestion Management for the ISDN Frame Relaying Bearer Service, International Telecommunication Union, Recommendation I.370, October 1991.

17. Generic System Requirements in Support of Switched Multi-Megabit Data Service, Bellcore, Document TR-TSV-000772, Issue 1, May 1991.

18. Report on the Broadband ISDN Protocols for Providing SMDS and Exchange Access SMDS, Bellcore, Document SR-NWT-002076, Issue 1, September 1991.

19. Support of Broadband Connectionless Data Service on BISDN, International Telecommunication Union, Recommendation I.364, March 1993.

20. Synchronous Frame Structures Used at 1544, 6312, 2048, 8488, and 44736 kbit/s Hierarchical Levels, International Telecommunication Union, Recommendation G.704, 1995.

21. Requirements for Private Branch Exchange Switching Equipment, ANSI/TIA/EIA-464-B, 1996.

22. Frame Relay/ATM PVC Network Interworking Implementation Agreement, Frame Relay Forum, Document Number FRF.5, December 1994.

23. Frame Relay/ATM PVC Service Interworking Implementation Agreement, Frame Relay Forum, Document Number FRF.8, December 1994.
24. DSS1 Signalling Specifications for Frame Mode Basic Call Control, International Telecommunication Union, Recommendation Q.933, 1992.
25. Frame Based User To Network Interface (FUNI) Specification v2.0, ATM Forum, Document AF-SAA-0088.000, July 1997.

CHAPTER 8

IP Over ATM

An important aspect in the acceptance of ATM as an infrastructure for data traffic is its ability to interwork with legacy network equipment and with the protocols used by these network components. The most important protocols currently in use are the Transmission Control Protocol (TCP) and the Internet Protocol (IP) [1].

We will begin this chapter with a discussion of ATM addressing among switching elements within an ATM network. Next, we will describe the Private Network Node Interface protocol for routing within ATM networks. We will then discuss how Network Layer protocols, such as IP, can operate over ATM networks.

In this chapter we will discuss the following network layer protocols:

- Classical IP over ATM
- Next Hop Routing Protocol
- LAN Emulation
- Multiprotocol over ATM

Finally, we will discuss procedures for the encapsulation of multiple protocols over ATM.

1. Addressing within ATM Networks

In an ATM network, the equivalent to an Ethernet MAC address is the ATM address. As such, when carrying IP traffic over ATM networks the first task is to resolve a destination IP address into an ATM address. Then a route must be established to forward the traffic to the destination. Since ATM is a connection-oriented technology, once the route is determined a connection must be established. If PVCs are used, then the connection has already been established and data transmission may begin immediately. But if SVCs are used, data transfer must be preceded by signalling procedures.

ATM addresses are referred to as Network Service Access Point (NSAP) addresses because they use a format similar to the OSI NSAP format [2]. This format calls for addresses to be specified in (network.host) form where the network parameter identifies a particular subnet while the host parameter identifies a particular end system on that subnet. An IP address is an example of an NSAP. ATM addresses, however, are not Network Layer addresses but are Data Link layer addresses. Such address-

Chapter 8: IP over ATM

es are more properly called Subnetwork Attachment Point (SNAP) Addresses that represent the unique addresses used by stations attached to a particular subnetwork. As such, an ATM address is more properly called an NSAP *format* address.

ATM addresses are defined to follow one of four formats as shown in Figure 8-1 [3]. These addresses consist of an Initial Domain Part (IDP), and a Domain Specific Part (DSP). The IDP identifies the authority responsible for allocating and assigning values of the DSP. The IDI consists of an Authority and Format Identifier (AFI), and an Initial Domain Identifier (IDI). The AFI identifies the ATM address format:

- Data Country Code (DCC) ATM Format: In this format, the IDI specifies a country code that identifies the country in which the address is registered. Country codes are specified in ISO 3166.

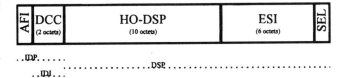

Figure 8-1a. DCC ATM address format.

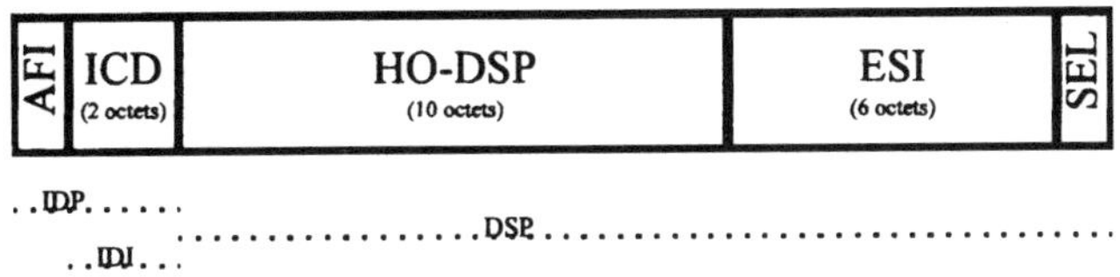

Figure 8-1b. ICD ATM address format.

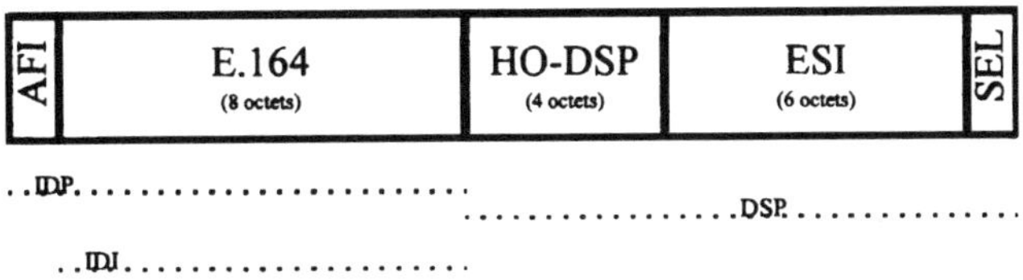

Figure 8-1c. E.164 ATM address format.

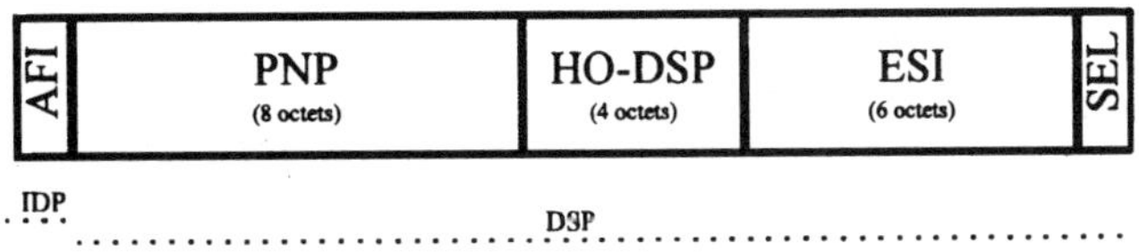

Figure 8-1d. PNP ATM address format.

- International Code Designator (ICD) ATM Format: In this case, the IDI is the code for an international organization. The British Standards Institute (London, England) is the registration authority responsible for assigning ICDs.
- E.164 ATM Format: The IDI in this format is an Integrated Services Digital Network (ISDN) number as specified in ITU Recommendation E.164. This format is generally used in public carrier networks to specify a telephone number.
- Private Numbering Plan (PNP) ATM Format: PNI supports proprietary and other nonstandard network addressing formats that do not conform to E.164 [16]. In the PNP format, the AFI is encoded to specify "Local IDI."

The DSP is subdivided into three fields:

- High Order DSP (HO-DSP): This is the "network address" portion of the NSAP format address and is specified by the authority identified in the IDI. This field should be constructed in such a manner to allow hierarchical addressing. Hierarchical addressing simplifies routing by allowing a group of end systems to appear as a single logical node to end systems outside of the group.
- End System Identifier (ESI): This field identifies a specific end system within the network. This field must be unique for a given combination of the IDI and HO-DSP fields. This field may be set to a globally unique identifier such as an IEEE MAC address.
- Selector (SEL): This field is not used for ATM routing but may be used by end systems.

2. Routing within ATM Networks

ATM addressing establishes a basis for the routing of paths between connection endpoints. Routing within ATM networks occurs in private as well as public networks by different procedures. Routing in public networks is not subject to standardization; the efficiency of one public carrier's routing procedures vis à vis another is often a basis for competitive differentiation between service offerings.

Within private networks the ATM Forum established a standard called the Private Network-Network Interface (PNNI) [4]. As shown in

Figure 8-2, PNNI is a hierarchical routing protocol that includes the following components:

- ATM Switching Nodes: identified by an ATM address.
- Peer Groups: a logical grouping of ATM nodes.
- Border Node: an ATM node that has a link to a node in another Peer Group.
- Horizontal Link: a logical link between ATM nodes within the same Peer Group.
- Outside Link: a logical link between ATM nodes within different Peer Groups.
- Peer Group Leader: an ATM node that is responsible for aggregation of peer group information.

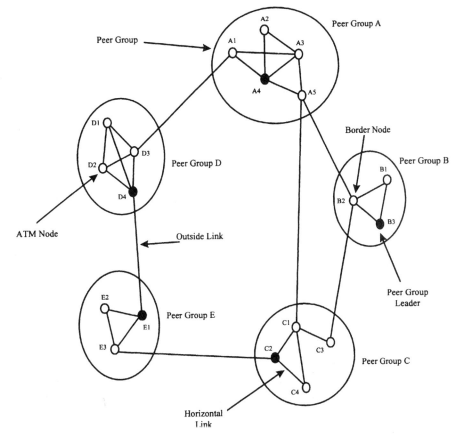

Figure 8-2.
PNNI hierarchy showing lowest level logical nodes.

The functions of the PNNI routing protocol include:

- Discovery of Neighbors and Link Status.
- Synchronization of Topology Databases.
- Flooding of PNNI Topology State Elements to other Peer Group members.
- Election of Peer Group Leaders
- Summarization of Topology State information.
- Construction of a Routing Hierarchy.

2.1 Discovery of Neighbors and Link Status

Figure 8-2 shows a logical configuration of ATM nodes that are supported by PNNI. The collection of nodes represents an ATM domain, the entire set of ATM nodes that are capable of communicating. Each node participating in the PNNI routing protocol must be configured with a Peer Group ID. This allows each node to recognize other nodes that are in the same peer group. At initialization (either of an ATM node, or a link) a node first discovers its neighboring nodes by executing the Hello Protocol. This involves sending a Hello Packet to any neighboring nodes that might be on the far end of the link. This packet is sent over a well-known VCC known as a PNNI Routing Control Channel (RCC). The Hello packet contains information that identifies the node to the neighbor. This information includes:

- Node ID: uniquely identifies the logical node. In the PNNI hierarchy a logical node is an abstraction. At the lowest level in the hierarchy, a node is an ATM switch. At a higher level, the node identifies a peer group, etc. The node ID is of significance in PNNI routing.
- ATM End System Address: this is the ATM address that is associated with the logical node identified in the Node ID field.
- Peer Group ID: identifies the peer group for this logical node.
- Port ID: this information associates a logical link with a physical port on an ATM switch.

Through the exchange of Hello packets, the neighboring nodes are able to determine their respective peer group affiliations. If the nodes are in the same peer group, they proceed to synchronize topology databases.

2.2 Synchronization of Topology Databases

A node's topology database represents that node's view of the ATM network domain. This database provides sufficient information to allow the node to compute a route to any ATM address reachable within the routing domain. Database synchronization is the exchange of information between neighbor nodes over a horizontal link resulting in nodes that have identical topology databases. This is accomplished by each node collecting its topology information into PNNI Topology State Packets (PTSP) that are comprised of one or more PNNI Topology State Elements (PTSE). PTSEs provide:

- Nodal Information: information about the node, its configuration, and its capabilities.
- Reachability Information: consists of a set of ATM addresses and logical nodes that are reachable from this node.

After completing database synchronization, the neighboring nodes involved in the exchange may advertise reachability to one another over the link in question.

2.3 Flooding of PTSEs

Flooding of PTSEs involves a node sending PTSPs to each neighboring node. This process allows all nodes within a peer group to update their topology databases with information on recent topology changes. Note that for an initializing ATM switch it would have received PTSEs from each neighbor during database synchronization. When a node receives a PTSP, it inspects each PTSE and sends an acknowledgment back to the sending neighbor. If a PTSE contains new information, it is installed in the node's topology database and flooded to all neighbors except the one from which it received the PTSE.

Flooding is an ongoing activity. PTSEs contained in topology databases are subject to aging and removal if not confirmed occasionally by PTSEs from neighboring nodes. A PTSE may only be updated by the node that originally provided the information.

2.4   Election of Peer Group Leader

PNNI supports a hierarchical node structure. Thus, individual nodes within Peer Group A may be represented by A1, A2, etc., while the collection of nodes in the peer group may be represented at a higher level by logical node A, as shown in Figure 8-3.

Flooding takes place only within a peer group. The Peer Group Leader (PGL) is responsible for aggregating information within the peer group and flooding that information to other PGLs. Thus, the PGL of Peer Group A, A4, is logically represented by Logical Node A in Figure 8-3. The ATM address for Logical Node A is the same as that of Node A4. The Peer Group Leader Election (PGLE) process is based upon information that was originally exchanged through Hello packets. Information about the elected PGL is subsequently flooded throughout the peer group to give all peer group members a consistent view of the PGL.

How does the PGL find out about the other peer groups? When border nodes execute the Hello protocol and receive a Hello packet from a neighbor with a different peer group ID, this informs the node that the neighbor is in a different peer group. This information is eventually flooded through the peer group, which ensures that the PGL will be informed of the existence of the outside link to another peer group.

The next step is that the PGL must establish an RCC with the PGL in the other peer group. For example, if nodes A5 and B2 in Figure 8-2 discover a border node relationship, node A floods information which associates the peer group node ID of Logical Node B with the ATM address of node B2. The PGL for Logical Node A (node A4) then establishes an SVC to B2 (through node A5) which is to serve as the RCC between peer group A and peer group B. Any subsequent PNNI information sent from logical node A to the logical node B peer will be sent from A4 to B2 (reachable through A5). B2 will then flood the information throughout peer group B.

In this example, we see that the only special function performed by the PGL is in determination of the information that is to be sent to the higher level peer group neighbors. Otherwise, the PGL has no special distinction within the peer group.

Chapter 8: IP over ATM

Figure 8-3.
Peer group level of the PNNI hierarchy.

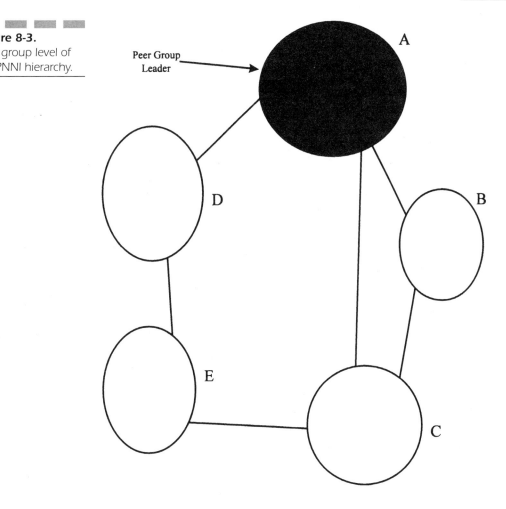

2.5 Summarization of Topology State Information

Were each PGL to simply broadcast a complete copy of the topology database seen within the peer group, this could quickly result in huge topology databases for routing domains of even modest size. This, in turn, would make PNNI a less scaleable protocol to large network size. Thus, the PGLs summarize peer group information prior to flooding to the higher-level peer group. For instance, the ATM addresses for each node represented by A1-A5 may typically have a common address prefix. Thus, rather than sending PTSEs for each ATM address, logical node A will send a single PTSE containing the address prefix common to the lower level peer group A nodes. This informs other peer groups that any address with the given prefix is reachable through peer group A.

2.6 Construction of the Routing Hierarchy

Following the exchanges described above, each node in the routing domain will have sufficient information to construct a route from any node to any other node. The hierarchical nature of PNNI allows this to be done without each node having detailed address maps for each individual ATM switch within the domain. When SVC connection establishment requests are made, the node will route a connection through the network to establish the requested connection to the addressed connection endpoint if the call is not rejected anywhere along the path to the requested endpoint. PNNI routing the originating ATM switch calculates the route through the network based upon its own topology database information.

When a PNNI-capable ATM switch receives a SETUP message across a UNI, it determines a route and then appends information elements (IEs) to the SETUP message, which specifies which PNNI nodes are to be transited along the path to the requested connection endpoint. This is referred to as a Designated Transit List (DTL) which contains a list of ATM addresses. The SETUP message is then routed to each of the identified nodes in the DTL. Since aggregated information may be used in this DTL if the route crosses outside links, it is the responsibility of each border node at the entry point of a peer group to reconcile the details of how the SETUP message is to be routed within the peer group. For instance, the SETUP message from a TE connected to A2 to a TE connected to E2 might contain a DTL with the following IEs at the originating switch:

- ATM address (A4)
- ATM address (A1)
- ATM address (D)
- ATM address (E)
- ATM address (E2)
- ATM address (Endpoint TE)

Chapter 8: IP over ATM

As each switch processes the SETUP message, it removes the IE containing its address from the DTL and passes the message on to the next node. Thus, at the entry to peer group D (node D2), the DTL might be as follows:

- ATM address (D)
- ATM address (E)
- ATM address (E2)
- ATM address (Endpoint TE)

Node D2 must then expand a route through peer group D to peer group E. Thus, it may modify the DTL to the following:

- ATM address (D3)
- ATM address (D4)
- ATM address (E)
- ATM address (E2)
- ATM address (Endpoint TE)

At the entry to peer group E (node E1), the DTL would be:

- ATM address (E)
- ATM address (E2)
- ATM address (Endpoint TE)

Since node E1 has a direct route to node E2, the SETUP message is forwarded to the destination ATM switch which ultimately forwards the SETUP message to the connection endpoint.

2.7 Route Selection

Selection of the "best" (among multiple) routes between two endpoints is determined within the PNNI routing protocol based upon two types of parameters:

1. Metrics: a set of parameters that applies to all nodes and links along a given path between endpoints. A metric is used to determine the suitability of the path based upon parameters whose values are meaningful when applied over the entire end-to-end path

2. Attributes: a set of parameters that are associated with a specific link or node. An attribute is used to determine whether a given link or node is suitable for inclusion in a route based upon parameters whose values may be associated with a specific link or node.

PNNI routing metrics include CDV and maximum CDT. PNNI routing attributes include CLR_0, and CLR_{0+1}, attributes which characterize the performance of a specific node. Other PNNI routing attributes include Maximum Cell Rate and Available Cell Rate, attributes that characterize a specific link. Recalling from our discussion of ATM traffic management in an earlier chapter we note that CDV, maximum CDT, CLR_0, and CLR_{0+1} were all parameters which could be derived from ATM OAM cells. This observation also helps us clarify the difference between metrics and attributes: CDV and maximum CDT were derived based upon the cumulative value derived over a path between the OAM source and destination points through the use of time-stamped OAM cells. CLR_0 and CLR_{0+1}, however, were values that applied specifically to the destination node which was the recipient of a performance-monitoring OAM cell.

These observations allow us to make the observation that built into PNNI routing is the ability to make routing decisions within ATM networks based upon QoS parameters.

3. Classical IP over ATM

Classical IP over ATM is described in RFC 2225 [5]. It allows IP traffic to be carried over ATM networks by resolving destination IP addresses into ATM addresses, which are then used for routing the traffic over the ATM network to the destination. The mechanism for performing this address resolution is called the ATM Address Resolution Protocol (ATMARP).

The basic components of a Classical IP over ATM implementation are:

- ATM-attached Stations: These may be End Stations, or LAN Access Devices. Groups of these stations are configured to form one or more Logical IP Subnets (LIS).
- One or more ATM Switches.

- ATMARP server(s): RFC 2225 does not specify where in the ATM network the ATMARP server is to reside but it is typically implemented in an ATM switch. Each LIS must have an ATMARP server which must be able to resolve IP addresses into ATM addresses for all hosts within the LIS. A single ATM switch may act as an ATMARP server for multiple LIS', but the switch must be a member of each LIS for which it is an ATMARP server. In a PVC-only ATM network, no ATMARP server is required.
- Router: Within classical IP over ATM, each LIS operates and communicates independently of other LIS' on the same ATM network. As such, hosts within an LIS may establish direct connections over the ATM network. However, when hosts on different LIS' wish to communicate, all traffic must be sent to a router, which forwards the traffic to the destination. The router must have membership in each LIS for which traffic forwarding service is to be performed.

That traffic crossing an LIS boundary must go through an intermediate router is a carryover from RFC 1122, which specifies requirements for communication between Internet hosts [6]. For this reason, this procedure is called "Classical." One of the reasons for using a "classical" model was security: routers provide firewalls that prevent unwanted access to an LIS. Another reason for using the "Classical" model is that it allows network administrators to configure ATM networks using the same models that they use in Legacy networks. An unfortunate result of use of intermediate routers when sending traffic between LIS' is that it prevents guaranteed end-to-end QoS.

Figure 8-4 shows an ATM network comprised of 3 LIS', where each LIS is defined by organizational structure: Engineering, Finance, and Marketing. Using this figure we are able to motivate discussion of the two types of communication in a Classical IP over ATM network: (a)communication between stations in the same LIS, and (b)communication between stations in different LIS'.

3.1 Communication between Stations in the Same LIS

In this case we have Station E1 communicating with Station E3 over the ATM network where both stations are in the Engineering Department

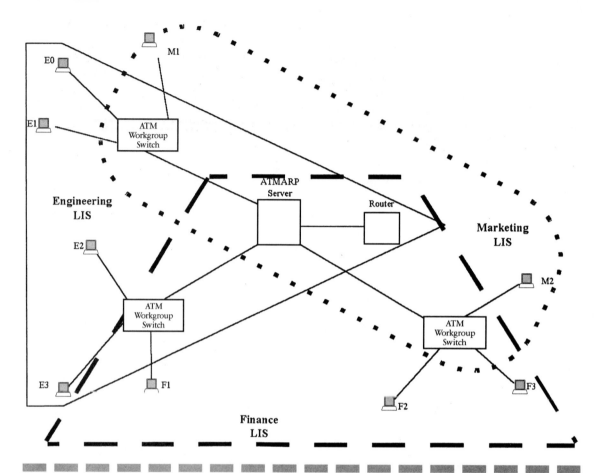

Figure 8-4. Logically independent IP subgroups on an ATM network.

LIS. E1 has an IP address to E3 but does not know the ATM address of E3. For SVC service the transaction proceeds as follows:

1. After unsuccessfully searching its IP to ATM address translation table for the ATM address of E3, E1 sends an ATMARP request to the ATMARP server for the LIS using the IP address for E3.

2. The destination IP address belongs to a station within the LIS so the ATMARP server for the LIS sends an ATMARP reply to E1 containing the ATM address of E3.

3. E1 adds an entry in its IP to ATM translation table for E3 based upon the contents of the ATMARP reply and then establishes a direct connection to E3 using ATM signalling.

4. E1 and E3 exchange data.

5. After the transaction has completed the connection is maintained for some period of time to allow subsequent traffic to be immediately exchanged between the two stations without the latency of another connection setup.

In the case of PVC service, if a PVC exists between stations E1 and E3, E1 would search its IP-to-ATM address table and find the ATM address of E3 and a VPI/VCI assignment for the PVC.

3.2 Communication between Stations in Different LIS'

In this example, we have Station E0 in the Engineering Department LIS, communicating with Station F2 in the Finance Department LIS. In this case the transaction (again, for SVC service) proceeds as follows:

1. After unsuccessfully searching its IP to ATM address translation table for the ATM address of F2, E0 sends an ATMARP request to the ATMARP server for the LIS using the IP address for F2.

2. The destination IP address belongs to a station in another LIS so this communication must go through a Router. The ATMARP server sends a reply to E1 containing the ATM address of the Router.

3. E0 adds an entry in its IP to ATM translation table for the Router based upon the contents of the ATMARP reply and then establishes a direct connection to the Router using ATM signalling.

4. E0 sends packets to the Router.

5. Assuming that the Router does not know the ATM address for F2, the Router issues an ATMARP request to the ATMARP server using the IP address of F2. Note that this time the request goes to the ATMARP server for the Finance LIS.

6. The destination IP address belongs to a station within the Finance Department LIS so the ATMARP server for the LIS sends an ATMARP reply to the Router containing the ATM address of F2.

7. The Router adds an entry in its IP to ATM translation table for F2 based upon the contents of the ATMARP reply and then establishes a direct connection to F2 using ATM signalling.

8. The Router forwards the packets to F2.

It is worth noting that in Figure 8-4, the Router has a single physical interface to the ATM network. Thus, an ATM device sends packets to the router, which then forwards the packet out over the same ATM interface. A router that receives and forwards packets over a single ATM interface is called a *one-armed router*. This is the recommended configuration as per RFC 2225.

There are differences between the operation of a broadcast-oriented network such as Ethernet and a Non-Broadcast Multiple Access (NBMA) network like ATM. For example, we see in the example above that address resolution replies are not broadcast to all stations in the LIS as would be the case on a broadcast network like Ethernet. Thus, ATM stations do not learn from other stations' address resolution queries and must issue their own requests.

Another difference that can also be seen in this example. In Ethernet networks, LIS' are defined by the number of stations that share a broadcast domain. Limitations upon the span of Ethernet segments means that LIS' in Ethernet are geographically defined. Stations on an ATM network are not restricted in this manner. Figure 8-4 shows stations on different ATM switches being defined to belong to the same LIS. Geographically, these stations could theoretically be anywhere. This logical association of workstations is called a Virtual LAN.

ATMARP packets under Classical IP over ATM are encoded in AAL5 Protocol Data Units (PDUs) using LLC/SNAP encapsulation as specified in RFC 1483 [7][8]. Classical IP over ATM is concerned only with the maintenance and resolution of IP and ATM addresses, and not with subsequent connection establishment between end stations, or encapsulation rules used for data transfer over the connection.

4. Next Hop Resolution Protocol

In the last section we saw that under Classical IP over ATM, when two end stations are in different LIS', traffic must travel over two IP hops: the first hop from the source to the router, and the second hop from the router to the destination. This two-hop route reduces performance over the network—maybe even needlessly so since the two stations, residing on the same NBMA network, could be connected directly via an SVC. Such a route, which crosses LIS boundaries is called a *cut-through* route.

The Next Hop Resolution Protocol (NHRP) supports cut-through routing on NBMA networks as shown in Figure 8-5 [9]. In NHRP, IP to

Chapter 8: IP over ATM

Figure 8-5.
NHRP resolution request/reply (single hop case).

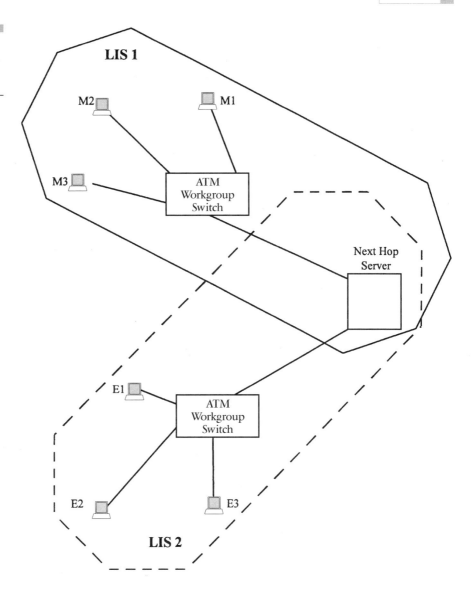

ATM address resolution queries are sent from a Next Hop Client (NHC) to a Next Hop Server (NHS) using an NHRP Resolution request. The NHS is able to respond to NHRP resolution requests for any set of addresses for which it is configured to do so, without regard to LIS boundaries as shown in Figure 8-5:

1. After unsuccessfully searching its IP to ATM address translation table for the ATM address of E2, M1 sends an NHRP Resolution

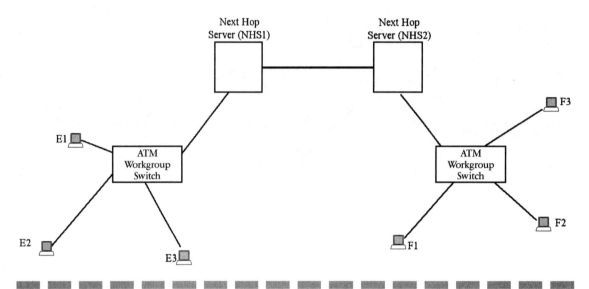

Figure 8-6. NHRP resolution request/reply (two hop case).

Request to the Next Hop Server for the LIS containing the IP address for E2.

2. Since the NHS knows the ATM address for E2 it sends an NHRP Resolution Reply to M3 containing the ATM address of E2.

3. M3 adds an entry in its IP to ATM translation table for E2 based upon the contents of the NHRP Resolution Reply and then establishes a direct connection to E2 using ATM signalling.

4. M3 and E2 exchange data.

In the event that the recipient NHS is not able to provide a reply, the request is forwarded to second NHS as shown in Figure 8-6:

1. E3 sends an NHRP Resolution Request to NHS1 requesting the ATM address of F1.

2. Since NHS1 does not know the ATM address of F1, it forwards the NHRP Resolution Request to NHS2.

3. Since NHS2 does know the ATM address for F1, it sends an NHRP Resolution Reply to E3 containing the ATM address of F1.

4. E3 establishes a direct connection to F1 using ATM signalling.

5. E3 and F1 exchange data.

If, in the example of Figure 8-6, NHS2 cannot resolve the destination IP address into an ATM address, the NHRP resolution request is sent to yet another NHS in a process that continues following an IP route to the destination until an NHS is found which can resolve the address. This is referred to as the *serving* NHS. How an NHS determines the next NHS to which an inquiry is to be sent for resolution is based upon Open Shortest Path First (OSPF) or similar IP routing algorithms [10].

Once the IP address is resolved to an ATM address by the serving NHS, an NHRP Resolution Reply is sent to the originating NHC. This request may be sent directly from the serving NHS to the originating NHC if a direct SVC/PVC exists. Alternatively, the reply may reverse the path followed by the NHRP request. Once the originating NHC receives the reply, it may then establish a direct SVC to the ATM address contained in the NHRP reply.

Returning to our example of communication between stations in different LIS' in Figure 8-4, an NHRP Resolution Request from station E0 would have generated a reply containing the ATM address of station F2. E0 would have then been able to establish an SVC directly to F2.

Security in the NHRP context is not defined by LIS boundaries, but is supported by the NHS. For instance, if a firewall is to be enforced between a source IP address and a destination IP address then the serving NHS would return the ATM address of a router. When the originating NHS establishes an SVC to the router, the router will then perform address filtering on the destination IP address, forwarding packets to the destination if the originator is authorized to send packets to that destination. Note that in this case, the router provides an effective firewall between source and destination ATM stations only so long as the originating station does not obtain the ATM address of the destination through other means.

A second scenario, shown in Figure 8-7, involves the use of a router in a multihop connection between an ATM station and a station that is not directly connected to the ATM network:

1. E3 sends an NHRP Resolution Request to the NHS requesting the ATM address of L1.
2. Since the NHS does not know the ATM address of L1, it forwards the NHRP Resolution Request to a Router that has advertised reachability to L1.

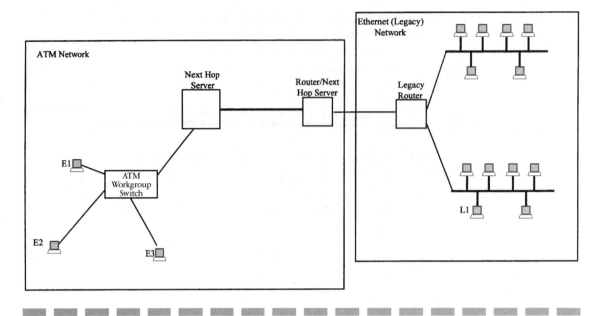

Figure 8-7. NHRP resolution request/reply (LAN-attached destination).

3. Since the Router knows that L1 is not on the ATM network, but that it is the closest exit point (from the ATM network) to L1, the router sends an NHRP Resolution Reply to E3 containing its own ATM address.

4. E3 establishes a direct connection to the Router using ATM signalling.

5. E3 and the Router exchange data.

6. The Router assembles the ATM cells into IP Packets and forwards them over the Legacy Network to a Legacy LAN Router.

7. The Legacy LAN Router forwards the IP Packets to L1.

NHRP can also be used to provide connectionless data transport. When sending an NHRP request, the originating NHC can send data packets while awaiting an NHRP reply. These packets would follow the same path as the NHRP request. Since each NHS receiving the NHRP request must form a contiguous hop-by-hop path to the destination, this mechanism guarantees that the data is being routed properly toward the destination. There is no guarantee, however, that this is the shortest path to the destination. For short-lived data flows, the data may be transmitted before the NHRP reply is received at the NHC. For longer-lived data flows, however, the NHC may establish a cut-through route once the reply is received. In this case, there is a transient period of time where there is traffic in transit on the network between the source and destination, which follow different routes. During this interregnum, it is possible that some ATM cells may arrive at the destination out of sequence.

5. LAN Emulation

As the name implies, the objective of LAN Emulation (LANE) is to give an ATM network appearance of a LAN by emulating the features and software interfaces of LANs [11]. LANE provides a MAC layer interface such as in Microsoft's Network Driver Interface Specification (NDIS), or Novell's Open Data-Link Interface (ODI). This relationship is shown in Figure 8-8. In the ATM host, LANE acts as a software shim between the NDIS/ODI interface and the AAL.

By offering the same MAC driver service primitives as traditional LANs, LANE is able to provide multiprotocol support for IP, IPX, AppleTalk, etc., without making any changes in existing applications. The obvious disadvantage of this approach is that individual applications are not able to take advantage of QoS features of ATM.

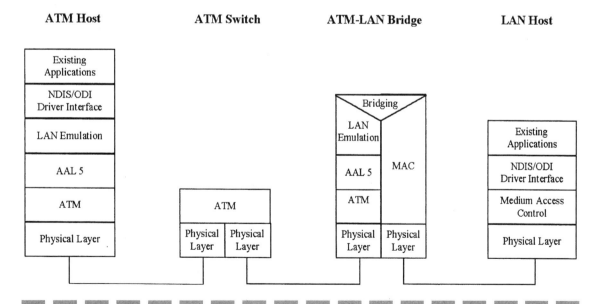

Figure 8-8. LAN emulation protocol stack.

LANE supports both Ethernet and Token Ring LAN topologies but does not support FDDI. Each emulated LAN is independent of all other LANs within an ATM network. There is no cut-through routing in LANE, users on one emulated LAN can only communicate with users on other emulated LANs through a router.

In a traditional LAN, a Network Layer address (such as an IP address) is resolved into a MAC address, which is broadcast over the LAN. LANE resolves a MAC address into an ATM address and then carries data, which would have been transported over a LAN, over a NBMA ATM network. LAN Emulation provides the appearance of a connectionless service to applications on the participating end stations.

The emulated LAN consists of the following components:

- LAN Emulation Clients (LEC): these are end systems on the ATM network that perform data forwarding, address resolution, and other control functions. The LEC could be either a workstation or a bridge.
- LAN Emulation Server (LES): The LES is responsible for responding to LAN Emulation ARP (LE ARP) requests from LECs that are members of the emulated LAN. LE ARP replies are based upon information at the LES that correlates MAC addresses and ATM addresses. Workstation LECs, which join the emulated LAN, register their MAC and ATM addresses with the LES. Proxy LECs, such as transparent bridges represent stations on a Legacy LAN to the emulated LAN by registering the station MAC addresses with the LES and associating them with the ATM address of the Proxy LEC.
- Broadcast and Unknown Server (BUS): The BUS provides broadcast/multicast services on the LANE. In addition, the BUS provides connectionless service by allowing an LEC to send data to a destination before the ATM address has been resolved and a direct SVC established.
- LAN Emulation Configuration Server (LECS): The LECS is responsible for assigning LECs to the appropriate LES. The LECS returns to the LEC, the ATM address of the LES associated with the assigned emulated LAN.

5.1 LANE Protocol Summary

Figure 8-9 shows how components communicate using the LAN Emulation protocol:

Figure 8-9.
Connections between LAN elements.

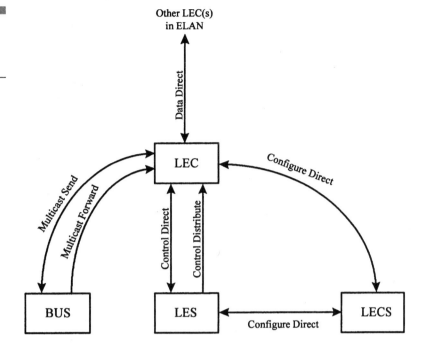

1. When an LEC attempts to join an Emulated LAN it establishes a Configure Direct connection with the LECS (the LEC may discover the LECS ATM address using ILMI procedures) to request configuration information, including the ATM address of the LES serving the requesting LEC [12]. The LECS returns the ATM address of the LES.

2. The LEC establishes a Control Direct connection with the LES. This connection is used for sending LE ARP requests from the LEC to the LES. This is a bidirectional connection that the LES may use (but is not required to use) to send LE ARP replies in response to LE ARP requests from the LEC. Upon successfully joining the Emulated LAN, the LEC may use this connection to register any number of MAC addresses (such as would be the case with a Proxy LEC), as well as Route Descriptors (for Source-Route Bridging) with the LES.

3. The LES establishes a Configuration Direct connection with the LECS to verify that the LEC is allowed to join the Emulated LAN. This is also a bidirectional connection over which the LECS returns a response informing the LES as to whether the LEC is or is not allowed to join the Emulated LAN.

4. If the LEC is allowed to join, the LES optionally establishes a unidirectional Control Distribute connection, which it may use to issue LE ARP requests to LECs. The Control Distribute connection is a point-to-multipoint connection to all LECs in the Emulated LAN. If this connection is not established then the LES uses the Control Direct connection for all control traffic to the LEC.

5. The LEC establishes a Multicast Send connection with the BUS. This connection is used to send broadcast and connectionless (unknown ATM address) traffic from the LEC. This is a bidirectional connection which the BUS may use (but is not required to use) to send broadcast and connectionless (from the standpoint of the originating LEC) traffic to the LEC.

6. The BUS optionally establishes a unidirectional Multicast Forward connection, which it may use to send broadcast and connectionless traffic to the LEC. This may be either a point-to-point connection between the BUS and each LEC in the Emulated LAN individually, or point-to-multipoint connection to all LECs in the Emulated LAN.

7. Data Direct connections are bidirectional point-to-point connections used by pairs of communicating LECs which are exchanging unicast traffic. As necessary, the LEC will use an existing PVC/SVC or will establish a new connection. If the ATM address for a given MAC address is not known to the LEC, it issues an LE ARP request to the LES. While awaiting an LE ARP reply, the LEC may direct data packets destined for the target LEC to the BUS using the Multicast Send connection. The BUS, in turn, forwards the packets over the Multicast Forward connection to all clients via flooding to ensure that the packet gets to its destination even if the destination is attached to a bridge that has not yet learned the destination's MAC address.

5.2 LANE Example

Figure 8-10 shows a LANE Scenario in which there are two emulated LANs. These emulated LANs provide the following types of connections:

- ATM Workstation to ATM Workstation
- ATM Workstation to Ethernet LAN Workstation (via IEEE 802.3 Bridge)

Chapter 8: IP over ATM

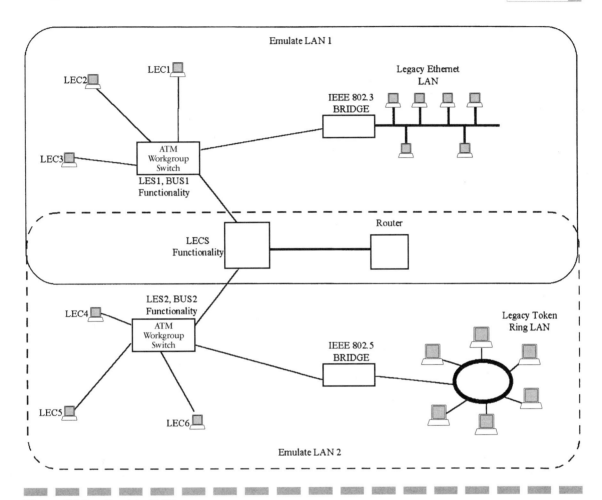

Figure 8-10. Emulated LANs.

- ATM Workstation to Token Ring LAN Workstation (via IEEE 802.5 Bridge)

It should be noted that LANE does not provide translational bridging between Ethernet and Token Ring LANs. Furthermore, communications that cross the emulated LAN boundary must go through the router.

Using Figure 8-10 we can construct an example of a transaction using a telnet session between two ATM hosts on the same IP subnet: in this case, ATM hosts LEC1 and LEC2. Telnet allows one host to connect to another using a command such as:

telnet *address*

where *address* is an IP address (or a label that resolves to an IP address). In a conventional LAN, an IP ARP packet is broadcast to all stations on the LIS to resolve the IP address to a MAC address. Once the MAC address is determined, the packet containing the telnet connection setup request is sent to the destination host.

Under LANE, the telnet application as well as all other protocols above the Network Layer (in this case the IP protocol layer and above) are unchanged. The differences are in how the IP ARP is resolved into a MAC address. The steps are as follows:

1. The IP ARP request is passed down the protocol stack to the LANE layer at station LEC1 where it is treated as a data packet to an unknown address. In response, LEC1 forwards the packet to BUS1 over the Multicast Send connection.

2. The BUS floods the LECs on Emulated LAN 1 (LEC1, LEC2, LEC3, and the IEEE 802.3 Bridge) with copies of the packet. Each recipient LEC passes the packet up through its respective protocol stack. In this case, the IP address corresponds to LEC2.

3. The IP ARP request contains the MAC address of LEC1, so LEC2 sends an LE_ARP request to LES1 to resolve the MAC address into the ATM address of LEC1. The LE_ARP request is sent from LEC2 to LES1 over the Control Direct connection.

4. LES1 sends an LE_ARP reply to LEC2 containing the ATM address of LEC1 over the Control Distribute connection.

5. LEC2 establishes a Data Direct connection and sends an IP ARP reply to LEC1.

Once LEC1 receives the IP ARP reply it is sent up the protocol stack to the process which originated the IP ARP request. This process can now send the telnet packet to the destination MAC address (LEC2) over the emulated LAN. Since Data Direct connections are bidirectional, the ATM connection will already be in place for LEC1 and LEC2 to communicate.

There is another significant difference between the operation of a Legacy LAN and an Emulated LAN here. In a Legacy LAN, all stations on the LIS "hear" IP ARP requests and replies and can thus learn the topology of the LIS as a shared experience. In the Emulated LAN, however, only the station issuing the LE_ARP request learns the MAC address to ATM address resolution.

6. Multicast IP Over ATM Networks

In addition to point-to-point connection between IP hosts, ATM also must be able to support IP multicast [13]. The solution supported in ATM Forum UNI 3.0/3.1 uses a Multicast Address Resolution Server (MARS) to enable multicast between IP hosts on the same LIS [14]. MARS is the multicast equivalent to the ATMARP server of Classical IP over ATM. A MARS responds to address inquiries using Class D IP addresses by returning a list of ATM addresses associated with a given multicast IP address.

6.1 Review of IP Multicasting

In IP version 4, addresses in the range:
 224.0.0.0 through 239.255.255.255
are termed "Class D" or "multicast group" addresses. These addresses abstractly represent the group of IP hosts that have decided to "join" the multicast group identified by the Class D IP address. The mapping between a Class D IP address to a hardware address of a specific host participating in the multicast group is performed by taking the 23 LSBs of the Class D IP address and mapping them into the 23 LSBs of the special Ethernet address:
 0x01005e000000
Membership in multicast groups uses procedures specified in the Internet Group Management Protocol (IGMP) as described in RFC 1112.

6.2 ATM Multicast IP Configuration

Class D IP groups may be supported in ATM by a mesh of directly connected IP hosts (Figure 8-11a), or by point-to-point connection from each host to a multicast server (MCS) which maintains a point-to-multipoint broadcast connection to each host (Figure 8-11b). In the latter case, the MCS may become a performance bottleneck limiting ATM multicast performance in comparison to the multicast mesh. The multicast mesh, however, requires a greater number of VCs.

Figure 8-11.
ATM multicast configurations.

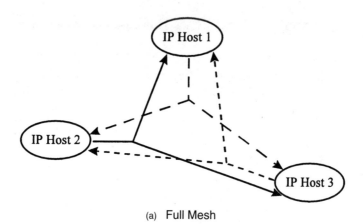

(a) Full Mesh

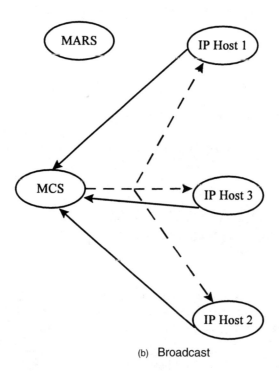

(b) Broadcast

The MARS and ATM hosts maintain a control communication relationship much like that the LES and LEC in LANE. The MARS communicates to the hosts over a point-to-multipoint cluster control VC (where a "cluster" is defined as a set of ATM hosts that participate in direct ATM connections to achieve multicasting among the group). This VC is used to inform end stations of changes in the multicast group—e.g., notification of when new members join the multicast group or when members leave. When an MCS is used, the MARS maintains a server control VC that is used to inform the server about current cluster membership. Individual ATM nodes communicate with the MARS by bidirectional point-to-point VCs. These are used for address resolution queries from the hosts, for receipt of MARS responses, to register membership in a specific cluster with the MARS, and to validate current cluster membership with the MARS.

The MARS plays no direct role in multicasting; its sole role is to maintain mappings between IP addresses and ATM addresses. The MARS maintains two types of address maps:

- Host Map, which contains a mapping between an IP address and a set of ATM addresses for each ATM host in the cluster.
- Server Map, which contains a mapping between an IP address and the ATM addresses for the MCSs associated with the cluster.

In a mesh configuration, when the Class D IP address has been resolved into the appropriate set of ATM addresses for the cluster, it is the responsibility of the host, to establish the appropriate point-to-multipoint VC (for the mesh configuration) to communicate with other cluster members.

In the case where an MCS is used, the host establishes a unidirectional point-to-point VC to the MCS. This VC allows the host to become a sending participant in the multicast cluster, while traffic from other cluster members is distributed from the MCS over a point-to-multipoint VC to all cluster members. When an MCS is used, hosts inquiring to the MARS for Class D IP address resolution receive the ATM address of the MCS in the MARS response, rather than the list of ATM addresses of cluster members.

As MARS clients register with the MARS their intention to join or leave a cluster, each host, or the MCS, associated with the cluster is informed of this event and are responsible for establishing or tearing down the appropriate VC. In the mesh configuration, each cluster member receives notification of joins and leaves from the MARS over the Cluster Control VC. In the multicast server configuration, the MARS notifies all MCSs in the cluster of joins and leaves over the Server Control VC.

7. Multiprotocol over ATM (MPOA)

MPOA integrates LANE and NHRP to provide cut-through routing between end stations on different emulated LANs. MPOA is based upon the notion of *virtual routing* in which the two functions performed by conventional routers:

- Route Computation, which involves the processing of routing protocols to determine the route for sending traffic from a source to a destination.
- Packet Forwarding, which uses the *source address* and *destination address* fields in the packet header, in addition to next hop information determined during route computation to forward packets to the next destination on the route.

They are split into two logical entities. Under MPOA there are centralized Route Servers, called MPOA Servers (MPS), which compute routes for a network of Packet Forwarders, called MPOA Clients (MPC). Unlike the IP-centric protocol discussed earlier in this chapter, MPOA is a Network, or Layer 3, protocol, which is, in principle, able to adapt multiple Network layer protocols (such as IPX, AppleTalk, etc.) to ATM networks.

As shown in Figure 8-12, an MPOA System consists of the following elements [15]:

- Edge devices, and ATM host devices, which have LAN Emulation Client, Packet Forwarding, and MPOA Client capabilities. An MPC may be associated with more than one LEC (such as a LAN bridge) but a given LEC may only be associated with a single MPC. There may also be multiple MPCs associated with a device (such as for multiprotocol devices).
- Routers, which have routing, LEC, Next Hop Server, and MPOA Server capabilities. There may be multiple MPSs associated with a router, and there may be multiple LECs associated with an MPS. A given LEC may be associated with only one MPS.
- Emulated LANs.

Chapter 8: IP over ATM

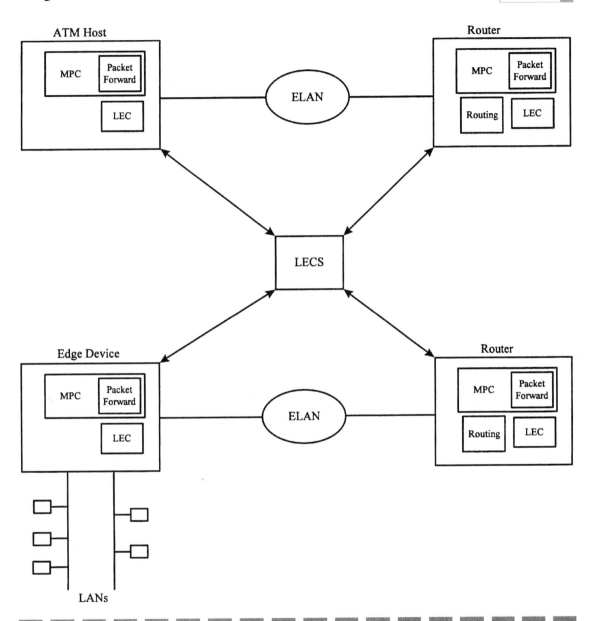

Figure 8-12. MPOA components.

- LAN Emulation Configuration Server: MPSs and MPCs communicate with the LECS to retrieve configuration information (e.g., internetworking protocols recognized by the MPS/MPC).

Within the MPOA system there are two roles for the MPC:

- Ingress role: in this role the MPC is at the point at which data flow enters the MPOA system. The observable action at this point is an MPC, which is forwarding packets over an ELAN to a router that contains an MPS.
- Egress role: in this role the MPC is at the point at which data flow exits the MPOA system. The observable action at this point is an MPC, which is receiving packets from another MPC which are to be forwarded to end stations or users.

For packets being sent between end stations on different ELANs over an MPOA system, the ingress MPC forwards packets over the ELAN to an MPS, which routes the packets over the appropriate ELAN to the egress MPC, which then forwards the packets to the destination. MPOA components discover each other using the LANE LE-ARP protocol.

For long-lived flows, the ingress MPC will attempt to establish a cut-through route to the egress MPC. The ingress MPC determines the ATM address of the egress MPC by sending an MPOA Resolution Request to the ingress-side MPS. When this MPS is able to resolve the MPOA Resolution Request, a reply is sent to the ingress MPC containing the ATM address of the egress device. The ingress MPC then establishes a direct connection to the egress MPC.

MPOA Resolution Requests and Replies between an MPC and an MPS are examples of MPOA *Control Flows*. Control flows may exist between any two MPOA components over an SVC established for the purpose of control information exchange. The address used for this exchange is referred to as a Control ATM Address. Packet forwarding over a cut-through route between MPCs is an MPOA *Data Flow*. The address used to establish the cut-through route is referred to as the Data ATM Address. The mechanisms for discovering Control and Data ATM addresses are different; an LEC retrieves the Control Address of its MPS from an LES by making an LE-ARP request, while the Data Address of an MPC is retrieved by making an MPOA Resolution Request to an MPS.

7.1 MPOA Data Flow

All ingress packets received at an ingress MPC are examined to determine whether they have the destination MAC address of an MPS. If the MPC is enabled to perform MPC functions on packets of the protocol type contained in the packet, the MPC inspects the network layer destination address within the packet. Combining this with the Control ATM address of the MPS, the MPC constructs a key which is used to retrieve an entry from its Ingress Cache for the packet flow. If no entry exists, one is created.

If present the entry contains the Destination ATM address which may be used to communicate with the destination over the cut-through route. If this is a new entry, this field is blank. If the egress cache does not contain destination address information, the ingress MPC sends an MPOA Resolution Request containing the network layer address of the destination to the ingress-side MPS. If the ingress MPS can resolve the request, it sends a reply to the ingress MPC containing the ATM address of the destination. If the ingress MPS cannot resolve the MPOA Resolution Request, it uses NHRP to forward the Resolution Request to the egress MPS. The egress MPS resolves the resolution requests and sends a reply back to the ingress MPS, which in turn forwards an MPOA Resolution Reply back to the ingress MPC.

For a destination that resides on a different ELAN, initial packets are initially forwarded from the ingress MPC to the MPS for delivery. Prior to routing packets to the egress MPC, the MPS imposes an Egress Cache entry for the packet flow at the egress MPC. This is called *Cache Imposition*. Cache imposition begins with a Cache Imposition Request being sent from the MPS to the egress MPC. There are two keys by which the egress cache entries are created/located: Network Layer Destination Address, and the Source/Destination ATM address pair. The egress cache entry identifies the LEC, which is the destination, and the Data Link Layer (DLL) header that is to be appended to packets received over the cut-through route prior to forwarding of the packet to the LEC. The latter field is set by the Cache Imposition Request from the MPS.

If the egress MPC is able to accept an SVC for the cut-through route, a positive Cache Imposition Reply is returned to the MPS. At this point, the egress MPC is ready to forward packets to the destination.

The ingress MPC maintains a count of the number of packets sent to the destination over this flow in the Ingress Cache. When a configured threshold number of packets have been sent to the destination, the ingress MPC attempts to establish a cut-through route.

If there is a destination ATM address in the ingress cache entry, an SVC is established using that address. If a VCC already exists, packets on this flow are immediately directed over the existing VCC. The ingress MPC strips the DLL header from any packets that are sent over the cut-through SVC. The egress MPC appends the appropriate DLL header prior to forwarding the packet to the destination.

7.2 MPOA Communication Scenarios

In this section we discuss the following connection scenarios:

1. Traffic from one MPOA Host to another MPOA Host, both within the same ELAN
2. Traffic from an MPOA Host to an Edge Device serving a Legacy Host, both in the same ELAN
3. Traffic from a LAN host to an MPOA Host via an Edge Device, MPOA Host, and Edge Device within the same ELAN
4. Traffic from one LAN host to another LAN Host via Edge Devices, both Edge Devices within the same ELAN
5. Traffic from one MPOA Host to another MPOA Host, MPOA Hosts within different ELANs
6. Traffic from a LAN host to an MPOA Host via an Edge Device, MPOA Host and Edge Device within different ELANs
7. Traffic from an MPOA Host to an Edge Device serving a LAN Host, Edge Device, and MPOA Host in different ELANs
8. Traffic from one LAN host to another LAN Host via Edge Devices, Edge Devices within different ELANs

Scenarios 1–4 are the trivial cases handled by LANE. The Scenarios 5–8 assume that the MPS is associated with each ELAN.

7.2.1 Case 5: MPOA Host (MPC 1) to MPOA Host (MPC 2), different ELANs

Assuming that there is no established cut-through route between MPC 1 and MPC 2, packets will initially be forwarded from MPC 1 to MPS via LANE. MPS will route the packets to MPC 2 also via LANE. After MPC 1 determines that the packet threshold for cut-through establishment has been reached, MPC 1 will send an MPOA Resolution Request to MPS requesting the ATM address of MPC 2. MPS will send a Cache Imposition Request to MPC 2. MPC 2 will provide a positive Cache Imposition Reply to acknowledge that it can accept a cut-through SVC for the packet flow. MPS then sends an MPOA Resolution Reply to MPC 1. MPC 1 establishes a cut-through SVC to MPC 2 and proceeds to forward packets over this VCC after removing the DLL header from the packets. MPC 2 appends the appropriate DLL header to the received packets and delivers the packets to the destination.

7.2.2 Case 6: LAN Host (LH 1) to MPOA Host (MPC 1), different ELANs

Host LH 1 broadcasts packets destined for MPC 1 over its LAN which are forwarded by an Edge Device (ED 1) to MPS via LANE. MPS routes the packets to MPC 1 also via LANE. Eventually, ED 1 will attempt a cut-through route to MPC 1. This procedure follows the discussion of Case 5.

After the cut-through route is established, packets from LH 1 continue to be broadcast over the LAN, ED 1 will strip the DLL header and forward the packets to MPC 1, where the appropriate DLL header will be appended and the packets delivered to the destination.

7.2.3 Case 7: MPOA Host to LAN Host, different ELANs

This case is the reverse of Case 6.

7.2.4 Case 8: LAN Host (LH 1) to LAN Host (LH 2), via Edge Devices in different ELANs

Host LH 1 broadcasts packets destined for LH 2 over its LAN which are forwarded by an Edge Device (ED 1) to MPS via LANE. MPS routes the packets to the edge device serving LH 2, ED 2 also via LANE. Eventually, ED 1 will attempt a cut-through route to ED 2.

After the cut-through route is established, packets from LH 1 continue to be broadcast over the LAN, ED 1 will strip the DLL header and forward the packets to ED 2 where the appropriate DLL header will be appended and the packets forwarded over the destination LAN to LH 2.

8. Multiprotocol Encapsulation over ATM

While the focus of this chapter has been the IP protocol, the use of ATM for network interconnection over local and wide area networks must also provide procedures for the carriage of multiple protocols. These protocols include routed protocols (such as IP, IPX, etc.), and bridged protocols (such as Ethernet, Token Ring, etc.). For ATM these procedures are specified in RFC 1483 using AAL 5 [7]. Two methods are described:

1. The first method involves the multiplexing of multiple protocols over a single ATM connection. In this method, the protocol of a carried PDU is identified by an IEEE 802.2 Local Link Control (LLC) header, which may be followed by an IEEE 802.1a SubNetwork Attachment Point (SNAP) header that tells the receiver what type of protocol encapsulation follows the header. This method is called *LLC/SNAP Encapsulation*.

2. The second method involves assigning an ATM connection for each protocol to be carried. Using this method, the higher layer protocol is able to determine the protocol being carried explicitly by the connection identifier. This method is referred to as *VC Based Multiplexing*.

The encapsulation method must be negotiated between the endpoint at connection establishment to ensure that the receiver of the PDUs

Chapter 8: IP over ATM

delivered by the AAL is able to properly interpret the received data. When using SVCs, the encapsulation is defined in the *Broadband Low-Layer Information* element in the SETUP message.

8.1 LLC/SNAP Encapsulation

Figure 8-13 shows the format of an LLC/SNAP encapsulated IP datagram. The PDU header is 8 octets and consists of:

- LLC header (3 octets).
- Organizationally Unique Identifier (OUI) (3 octets). The OUI indicates the organization that administers the meaning of the Protocol Identifier field.
- Protocol Identifier (PID) (2 octets). In the case of IP, and other protocols not defined by the International Organization for Standardization (ISO), the PID is an Ethertype whose values are defined in RFC 1700 [8][17]. Ethertype 0x0800 indicates an IP PDU.

8.2 VC Based Multiplexing

VC multiplexing does not require a header for protocol identification. For example, an IP datagram may be carried over an ATM connection without the 8-octet header used in LLC/SNAP Encapsulation.

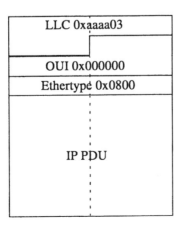

Figure 8-13. LLC/SNAP encapsulated IP PDU.

9. Interaction between ATM Routing and IP Routing

As described in this chapter, the IP over ATM methods are based upon an overlay model in which the IP protocol and underlying ATM protocol operate independently of one another. This can result in an IP routing protocol producing different routes than those produced by a PNNI computed route.

OSPF, for example, divides a network of routers into areas (similar to the peer groups of PNNI). The maximum number of routers within a given area is typically recommended by router vendors. Among the OSPF areas is a special area—Area 0, also known as the OSPF Backbone. All traffic between OSPF areas must be routed through Area 0. Certain routers are members of multiple areas (including Area 0), these are known as Area Border Routers (ABRs). ABRs summarize information within an area similar to the role served by PNNI PGL.

For routers with interfaces on an ATM network, OSPF could produce multiple hop routes where a single hop route might be computed using PNNI if the destination ATM address were known at the originating node.

NHRP allows a router to issue an address resolution request, which produces the shortest possible route through the ATM network. Thus, a NHRP resolution request produces a one-hop route if both routers are on the same ATM network. In the case where the destination router is not on the ATM network, NHRP produces a one-hop route to the nearest router to that destination router (in this case, the "nearest" router is determined by OSPF routing metrics).

References

1. Comer, Douglas, Internetworking with TCP/IP 2d Ed., Prentice-Hall, 1991.

2. Information Technology—Open System Interconnection—Network Service Definition, International Telecommunion Union, Recommendation X.213, November 1995.

3. ATM User-Network Interface Specification, Version 3.1, ATM Forum, September 1994.

4. Network-Network Interface Specification Version 1.0 (PNNI 1.0), ATM Forum, Document af-pnni-0055.000, March 1996.

5. Laubach, Mark, and Joel Halpern, Classical IP and ARP over ATM, Internet Engineering Task Force, RFC 2225, April 1998.

6. Braden, Ralph, Requirements for Internet Hosts—Communication Layers, Internet Engineering Task Force, RFC 1122, October 1989.

7. Heinanen, Juha, Multiprotocol Encapsulation over ATM Adaptation Layer 5, Internet Engineering Task Force, RFC 1483, July 1993.

8. Reynolds, Joyce, and Jon Postel, Assigned Numbers, Internet Engineering Task Force, RFC 1700, October 1994.

9. Luciani, James, Katz, Dave, Piscitello, David, Cole, Bruce, and Naganand Doraswamy, NBMA Next Hop Resolution Protocol, Internet Engineering Task Force, RFC 2332, April 1998.

10. Moy, John, OSPF Version 2, Internet Engineering Task Force, RFC 1583, March 1994.

11. LAN Emulation over ATM Version 1.0, ATM Forum, Dcoument af-lane-0021.000, January 1995.

12. Integrated Local Management Interface (ILMI) Specification Version 4.0, ATM Forum, Document af-ilmi-0065.000, September 1996.

13. Deering, Steve, Host Extensions for IP Multicasting, Internet Engineering Task Force, RFC 1112, August 1989.

14. Grenville Armitage, Support for Multicast over UNI 3.0/3.1 Based ATM Networks, Internet Engineering Task Force, RFC 2022, November 1996.

15. Multiprotocol over ATM Version 1.0, ATM Forum, Document af-mpoa-0087.000, July 1997.

16. Information Technology—Telecommunications and Information Exchange Between Systems Numbering and Subaddressing in Private Integrated Services Networks, ISO/IEC 11571, 1994.

17. Information Technology—Protocol Identification in the Network Layer, ISO/IEC Technical Report 9577, 1996.

# CHAPTER 9

TCP Over ATM

1. Introduction

The result of research funded by the Defense Advanced Research Projects Agency (DARPA), the Transmission Control Protocol, and Internet Protocol Suite, commonly referred to as TCP/IP has demonstrated its viability on a large scale. It forms the basic technology for the Internet. TCP provides a connection-oriented, flow-controlled, block data transfer protocol for the transmission of files and data streams.

TCP guarantees complete, error-free, sequenced data delivery over a "best effort" datagram delivery service but makes no guarantee regarding delay. TCP uses a sliding window to manage data flow, but does not use explicit rate control and has no explicit congestion notification.

But while TCP is "block oriented" and not "rate oriented," the protocol does adjust its data rate to available bandwidth and recovers from data loss and out-of-sequence delivery. It is the interaction between the flow control mechanisms of TCP and the traffic control mechanisms of ATM that can significantly impact the performance of TCP applications over ATM networks. In one study, relatively small cell losses of 0.03% to 0.04% reduced TCP throughput by 70% [1].

2. TCP Features

2.1 Sliding Window Flow Control

The TCP sliding window protocol allows the window size to vary over time. Each acknowledgment specifies the number of octets that have been received and also includes a window advertisement that indicates how many more octets of data the receiver is ready to accept. If the window advertisement indicates a larger window size than the previous acknowledgment, the sender increases the size of its sliding window in response. Likewise, when the window advertisement indicates a smaller window, the sender reduces its window accordingly.

The sliding window protocol provides flow control as well as reliable data transfer. The variable aspect of the window size allows TCP to adapt to the available bandwidth on the network. However, flow control based upon window advertisement is asserted only between communicating endpoints.

Without a mechanism for assertion of flow control between the source and intermediate nodes, it is possible for intermediate nodes to be overwhelmed by the amount of data being received by sources sending TCP traffic over the network. This condition is called congestion and is addressed by the congestion control mechanisms of TCP.

One of two things happens to data under congestion conditions within the network:

1. Data transit time across the network increases as intermediate nodes must queue received data in buffers;
2. As the buffers overflow, data is lost.

TCP maintains estimates of the round trip time (RTT), which measures the elapsed time between when data is sent from sender to receiver, and when acknowledgment of receipt of that data is received by the sender.[1] The RTT is the basis for determining a time-out period. If an acknowledgment is not received within the time-out period, packet loss is assumed and data is resent. In addition, the RTT value is updated to more accurately reflect conditions within the network.

If previously sent data was lost, this will be indicated in the received acknowledgment. In response, data is resent.

In the event of either time-out or lost data, the sending TCP entity responds to network congestion by reducing the window size. This has the effect of throttling back the data rate on the link.

During network congestion, neither TCP endpoint knows the location of congestion within the network, only that there is congestion somewhere inside the network.

2.2 TCP Window Size

In order to obtain maximum performance over a network, the "bit pipe" must be kept full, otherwise bandwidth which could be used to increase communication throughput is wasted. Under these circumstances, TCP becomes a "stop and wait" protocol, sending a window of data and then

---

[1] This is not a trivial estimate as TCP must be able to accommodate RTTs for communication paths that range from stations on a common LAN which are separated by a few feet, to stations separated by thousands of miles [2].

waiting for acknowledgments of previously sent data before sending additional data. The bit pipe is kept full if the TCP window size is large enough to continuously transmit data for a period of time equal to the RTT [3]. After RTT amount of time, the sender should receive an acknowledgment that would allow the sender to continue sending data without interruption. The minimum TCP window size (in octets) meeting this condition is determined by the bandwidth-delay product:

$$\text{TCP Window Size} \geq \text{Link Bandwidth} \times \text{RTT}$$

For TCP window sizes that are equal to the bandwidth-delay product, the window is able to keep the bit pipe full without packet loss. For TCP window sizes greater than the bandwidth-delay product, there must be sufficient capacity among the queues within network routing/switching nodes to absorb the excess packet traffic. The upper limit on the TCP window size with no packet loss is:

$$\text{TCP Window Size} \leq \text{Link Bandwidth} \times \text{RTT} + \text{Queue}$$

Since TCP fairly shares bandwidth among competing sessions, if all sessions have similar bandwidth-delay products, the total aggregate window size over all the sessions is the same as that specified above. For N simultaneous TCP sessions, each session will receive 1/N of the available bandwidth and requires a maximum TCP window size of (Link Bandwidth $\times$ RTT + Queue)/N.

The original TCP specification defined a maximum window size of 64Kbytes [4]. However, as shown in Table 9-1, broadband link rates produce bandwidth-delay products in excess of this limit. Extensions to TCP as defined in RFC 1323 allow much larger window sizes [5]. Thus to fully utilize the bandwidth on DS-3, and faster, links requires TCP implementations which include RFC 1323 extensions.

2.3 TCP Segment Size

The unit of transfer in TCP communications is called a *segment*, which consists of a header, followed by data. While not all segments need be of the same size, sending and receiving parties must agree upon a TCP Maximum Segment Size (MSS) that is to be used. In addition, the network also has a Maximum Transfer Unit (MTU) which defines the maximum IP frame size (in octets) which is transmitted over a given network.

TABLE 9-1.

Bandwidth delay products

Data Rate	Link Bandwidth (Mbps)	RTT (msec)	Bandwidth-Delay Product (kB)
T-1 Short Haul	1.5	20	3.8
T-1 Cross Continental	1.5	70	13.5
T-3 Short Haul	44	20	92
T-3 Cross Continental	44	70	390
OC-3 Short Haul	150	20	375
OC-3 Cross Continental	150	70	1,300

The TCP MSS is typically set to a value less than the IP MTU size to avoid fragmentation of the TCP segment.

Many TCP implementations set the MSS to be some multiple of 512 octets to ensure that the segment length is less than the default MTU value of 576 octets.[2] However, larger values of MSS allow TCP to send more data in fewer packets. This allows routers to operate more efficiently. Table 9-2 shows IP MTU and maximum TCP MSS for different networks. Note that SMDS and ATM networks support MTU/MSS sizes that are greater than six times those of Ethernet networks.

In ATM networks it should be noted that the MTU defines the maximum size of the AAL PDU that is passed by the IP protocol layer across the AAL-SAP. Within ATM, TCP segments must be fragmented into ATM cells. Since TCP has no knowledge of an underlying ATM network, cell loss exacerbates packet loss as loss of a single cell that is the same as loss of an entire packet. Given a nonzero CLR, increasing MTU size increases the probability of packet loss over ATM networks.

2.4 TCP Behavior

To motivate our discussion of the performance of TCP over ATM networks, we start with a discussion of the bandwidth-seeking behavior of

---

[2] This MTU default was used to limit the load on intermediate routers. However, many of today's routers are capable of transferring larger MTUs without fragmentation. The MTU path discovery option allows routers to determine the appropriate MTU size [6].

TABLE 9-2.

MTU/MSS sizes

Network	IP MTU	Maximum TCP MSS
Default	576	536
Ethernet	1,500	1,460
FDDI	4,352	4,312
SMDS/ATM	9,180	9,140

TCP. Not all TCP versions are the same, but more recent TCP implementations (Tahoe TCP and later) incorporate the following mechanisms:

1. Slow Start
2. Congestion Avoidance
3. Fast Retransmit

2.4.1 Slow Start

TCP has no means of determining the available bandwidth on a link. Thus, it must determine the amount of available bandwidth by trial and error. The objective of the slow-start algorithm is to gradually increase the TCP window size, and correspondingly, the bandwidth utilized over the link [7]. Slow start begins by setting a state variable representing the TCP window size, *cwnd* (for congestion window), equal to one TCP segment. For each acknowledgment received by the sender:

$$\text{cwnd}_i = \min(\text{cwnd}_{i-1} + 1, \text{WA}_{rcv})$$

where WA_{rcv} is the value of the window advertisement from the receiver.

If we assume that the TCP window size at the sender is not constrained by window advertisement from the receiver, then cwnd increases by 1 for each received acknowledgment. At this rate the TCP window size increases exponentially (in powers of 2) for each round trip time.

Thus, the "slow start" algorithm is not really very slow at all. The algorithm converges upon the maximum TCP window size, W_{MAX}, in time $RTT \times \log_2(W_{MAX})$.

2.4.2 Congestion Avoidance

For a large enough W_{MAX} there will eventually be packet loss. One mechanism for indicating potential packet loss is time-out while waiting for an acknowledgment. When this occurs, TCP enters congestion avoidance by setting $cwnd_i$ to half of $cwnd_{i-1}$. Thus, slow start ensures that data will be transmitted at least half the rate possible on the link. Since packet loss is an indication of congestion in the network, TCP does not exacerbate the problem by continuing exponential growth in TCP window size upon receipt of subsequent acknowledgements. In the congestion avoidance phase:

$$cwnd_i = cwnd_{i-1} + \delta$$

The value δ is chosen such that cwnd increases by 1 segment every RTT. Since (in the absence of packet loss) there are cwnd acknowledgments every RTT, δ = 1/cwnd segment. Assuming that the TCP implementation does not attempt to send partial segments, with congestion avoidance, W_{MAX} grows linearly with respect to time.

2.4.3 Fast Retransmit

Another mechanism for detecting packet loss is indicated in the acknowledgments, which are received by the sender. As we saw in our discussion of SSCOP, in a sliding window protocol, the acknowledgment from the receiver indicates the sequence number of the next expected packet. Thus, if a packet is lost, the receiver continues to indicate that packet as being the next expected packet even if later packets arrive first. The sender, then, sees duplicate acknowledgments. The ability of the fast retransmit algorithm to prevent timeout (and subsequent slow start) depends upon retransmission of the lost packet such that it is resent, and an acknowledgment received before a time-out occurs. Thus, for a time-out of N RTTs, the fast retransmit decision must be made within $(N - 1)$ RTTs from the time that the packet was originally sent. For reasonable time-out values, duplicate acknowledgements can be received (and trigger a fast retransmit response) before a timeout occurs for any transmitted packets that have not been acknowledged. Fast retransmit is triggered upon receipt of some number (typically three) of duplicate acknowledgments and resends the lost data. Three duplicate acknowledgments can typically be received in approximately 1.5 RTTs. Since this is also an indication of congestion in the network, fast retransmit also reduces the TCP transmit window to half of its current size and invokes congestion avoidance. Upon receipt of the missing data, the receiver acknowledges the total received data.

Fast retransmit requires that the receiver acknowledge receipt of every received packet rather than wait to piggyback an acknowledgment on data being sent in the reverse direction. If another packet loss is detected within the same window, fast retransmit will not be able to respond to a second packet loss quickly enough to prevent the TCP sender from timing out. Consequently, TCP reenters the slow start regime. This time, however, TCP maintains two state variables:

1. cwnd: set equal to 1
2. sst: the slow start threshold—set equal to half of the window size at the time slow start was invoked.

Slow start increases the window size exponentially from cwnd = 1 until packet loss is detected, or the window size is equal to sst. At this point congestion avoidance is invoked and the window grows linearly. As long as no more than one packet is lost within any window, TCP remains in congestion avoidance.

2.4.4 TCP Performance

Note that slow start is a condition to be avoided if possible. The impact of slow start is that the bit pipe is emptied as TCP becomes a send and wait protocol with a single segment in the bit pipe. This has a dramatic negative impact upon TCP performance. On the other hand, when congestion avoidance and fast retransmit operate, the bit pipe remains mostly full. This is the TCP regime that gives the best performance.

Another important contributor to TCP performance is TCP window size. TCP implementations that are limited to a 64kByte TCP window, and do not support RFC 1323 enhancements, can achieve at most 7.3 Mbits/sec transfer rates on a 70-ms cross-continental link. For a cross-continental DS-3 ATM link with a payload capacity of 37 Mbits/sec, this amounts to less than 20% link utilization.

3. TCP Performance Over ATM Networks

Because of the nature of TCP/IP traffic, UBR and ABR are the ATM QoS classes most frequently studied for this application. AAL Type 5 is the ATM adaptation method used.

Chapter 9: TCP OVER ATM

3.1 QoS and TCP Performance

While TCP and ATM QoS mechanisms operate independently, the performance of TCP is strongly influenced by the interaction of these mechanisms. TCP can run on an ATM-attached host, or on a host attached to a legacy network that interconnects to an ATM network through a router serving as an edge device. In either case, with UBR, the feedback control mechanism is as shown in Figure 9-1. Packet flow is controlled between communication endpoints based upon TCP mechanisms. UBR provides no traffic management feedback, with network congestion being indicated by lost or excessively delayed acknowledgments that are detected by the TCP source only. From the perspective of the ATM network, congestion may occur at the edge device or within the ATM network.

With ABR, however, there is an ABR control loop operating within the TCP control loop as shown in Figure 9-2. Thus, ABR pushes congestion outside of the ATM network to the edge devices.

3.2 Cell Dropping Strategies and TCP Performance

Another factor that influences TCP performance is the strategy used in selecting cells to drop within a congested ATM network. These strategies fall under the following categories:

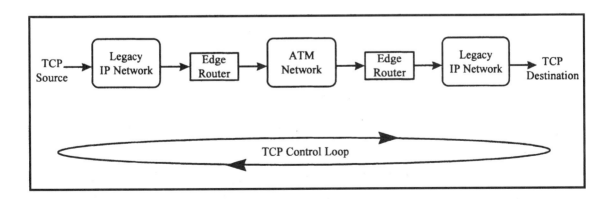

Figure 9-1. Flow control loop: TCP with UBR Qos.

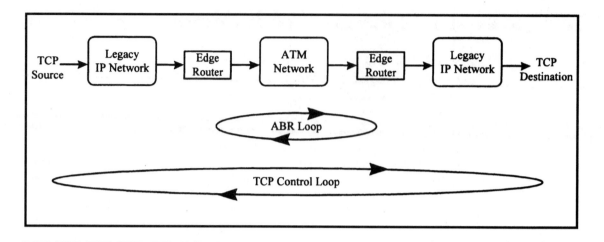

Figure 9-2. Flow control loop: TCP with ABR Qos.

1. Plain ATM
2. Partial Packet Discard
3. Early Packet Discard

3.2.1 Plain ATM

In a congested ATM network, "plain" ATM results in individual cells being dropped without regard to other cells. Thus, a single dropped cell results in a corrupted packet that triggers retransmission of the entire packet. As shown in Table 9-3, for a nonzero CLR, the larger the MTU size, the greater the probability of packet loss. Subsequent cells associated with a packet that has lost a cell then become "dead cells," which will be discarded at the destination. Transmission of such dead cells represents waste of ATM network resources that could be allocated to traffic, which contributes to the effective TCP throughput, or "goodput."

3.2.2 Partial Packet Discard

Partial Packet Discard (PPD) eliminates the transport of dead cells by dropping all but the last cell of a packet once a cell loss has occurred. PPD detects the last cell in a packet by inspecting the AUU field in the

ATM cell header, which provides an indication of the last cell in an AAL 5 SDU. Implementation of PPD requires that the ATM switch maintain per-VC state information to determine which VCs are using AAL 5 and PPD. The switch must also keep track of whether a cell has been dropped in the current AAL SDU.

It is important that the last cell not be dropped because that allows the AAL entity in the receiving endpoint to delineate one AAL 5 SDU from another. Should this cell be dropped, then a corrupted packet will be merged with a potentially uncorrupted packet resulting in retransmission of both packets. In addition, this would trigger a TCP slow start degrading TCP performance even further.

3.2.3 Early Packet Discard

While PPD represents an improvement over "plain" ATM, it still transmits cells within a packet that were sent prior to the cell that was dropped later due to congestion. Thus, even though these cells were initially "good" cells, they became "dead" cells as a result of a subsequent cell drop event. Early Packet Discard (EPD) attempts to anticipate in advance which packets will be successfully transferred in their entirety. Cells within packets that are expected to be transferred without loss are transmitted. All other cells are discarded by the ATM switch to enhance the probability of the "predicted" packets being successfully transferred across the ATM network.

EPD monitors buffer utilization in the ATM switch until buffer utilization reaches a defined threshold. Beyond that point, cells from "in progress" packets continue to be transmitted, but cells from all "new" packets are dropped.

EPD addresses the problem of unnecessary retransmission of cells, but with too conservative of a buffer threshold level, EPD may drop cells too aggressively resulting in packets being rejected that might have been successfully transmitted with a less stringent threshold.

TABLE 9-3. Cells/Packet

Packet Size (octets)	ATM Cells/Packet
512	11
1,500	32
4,352	91
9,180	192

3.2.4 Limitations of PPD and EPD

PPD and EPD operate on a per switch node basis without cooperation among ATM switches. Thus, there is still the possibility that resources within the ATM network are wasted transmitting "dead" cells, resources that upstream ATM nodes would have used transferring cells to the node at which the cell drop event occurs.

PPD and EPD operate with the UBR QoS class, which does not provide a guarantee with regard to CLR. Since ABR manages per-VC bandwidth allocation in response to congestion conditions within the network, PPD and EPD are not necessary.

3.3 TCP Over ATM Performance: Empirical Results

In 1995, a study performed on the Swedish University Network (SUNET) resulted in a series of experiments with TCP/IP performance over ATM wide area networks [1]. These experiments were performed during the early stages of ATM deployment using early generation ATM switching equipment. ATM service consisted of a CBR PVC, which could be considered as a serial line for the purposes of the experiments. Packet sizes used in the experiment were up to 9,180 octets.

In one experiment, aggregate throughput fell to approximately 30% of the expected rate. The reason was ultimately found to be due to a problem in a local ATM switch used in the experiment, which produced cell loss at the switch output port. The cell loss reported by the switch, however, was in the range 0.03% to 0.04%.

Cell loss in the range reported by this result suggests a CLR of 1 cell in every 2,500 to 3,300 cells. For a 9,180 octet packet (192 cells), and a uniformly distributed cell loss, this corresponds to one corrupted packet out of every 13 to 18 cells transferred.

Other experiments that were performed with traffic shaping at the traffic source being intentionally crafted to produce noncompliant traffic also showed that small cell losses within the ATM network (due to UPC enforcement) produced dramatic reductions in throughput. In one experiment using a UDP traffic source, a 0.8% cell loss produced a 90% reduction in UDP throughput. When cell loss had reached 4%, UDP throughput reduction was over 99%.

3.4 Simulation of TCP Over ATM Performance

There have been a number of simulation studies to investigate the factors influencing the performance of TCP over ATM [8][9][10][11].

3.4.1 1 TCP Over ATM in a LAN Environment

In [8] a configuration with 10 TCP senders directly connected to an ATM switch that sends TCP traffic to a single receiver, as shown in Figure 9-3, all links were chosen to be the same rate (approximately 141 Mbps) to test the performance of the bottleneck link between the ATM switch and the receiver. This study simulated TCP over ATM in a LAN environment. Thus, round trip delays in the study [8] were extremely short (12 µs), which produced a bandwidth-delay product of only 4 cells.

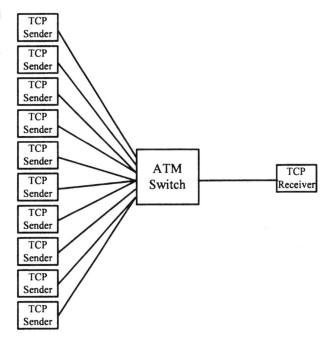

Figure 9-3.
Simulation scenario 1.

Simulation results were produced for plain ATM, PPD, and EPD with packet sizes varying between 512 to 9,180 octets and with output buffer sizes in the ATM switch varying between 256 to 8,000 cells. Packet transmissions were 15 seconds each in duration.

Plain ATM produced the poorest results overall with performance decreasing with increasing packet size as shown in Figure 9-4. At an out-

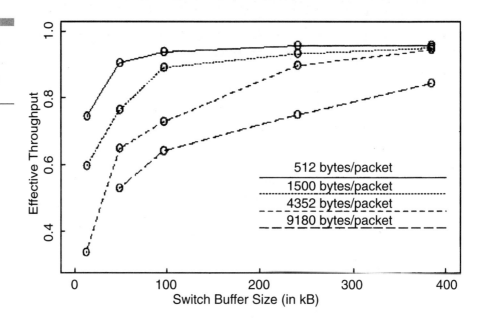

Figure 9-4.
10 TCP senders, 1 TCP receiver, 64kB window size, plain ATM.

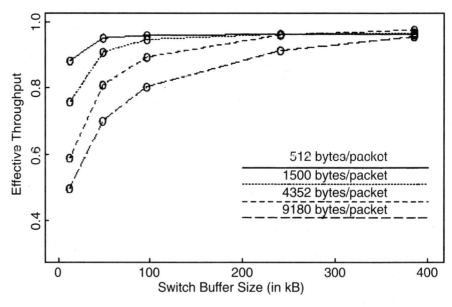

Figure 9-5.
10 TCP senders, 1 TCP receiver, 64kB window size, partial packet discard.

Chapter 9: TCP OVER ATM

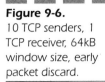

Figure 9-6.
10 TCP senders, 1 TCP receiver, 64kB window size, early packet discard.

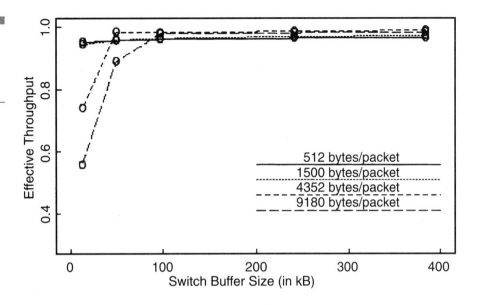

put buffer size of 256 cells effective throughput was 0% for some connections. As shown in Figures 9-5 and 9-6, the best overall results were produced using EPD. The EPD threshold for this study was set to 50% of output buffer capacity. Since EPD is biased to accept cells from "in-progress" packets, the slightly higher effective throughput for larger packets in Figure 9-6 may be due to the fact that larger packets have a better chance of being "in progress" when the EPD threshold is reached.

In [8], a baseline simulation was performed using a packet switch instead of an ATM switch. Output buffer management used a strategy in which packets were dropped from the tail of the buffer under conditions of congestion. As can be seen in Figure 9-7, the packet switch performance exceeded that of any of the cases using an ATM switch. This result is not surprising, since a packet switch is able to discard entire packets and therefore does not waste resources on delivery of partial packets, which must be retransmitted anyway.

3.4.2 TCP over ATM in a Short Haul Environment

In [9] the simulated configuration used 10 TCP senders and 10 TCP receivers as shown in Figure 9-8. Each sender and each receiver was connected to a separate edge device. Edge devices on the sending and receiving sides are connected to separate ATM switches with the switches

Figure 9-7.
10 TCP senders, 1 TCP receiver, 64kB window size, packet TCP.

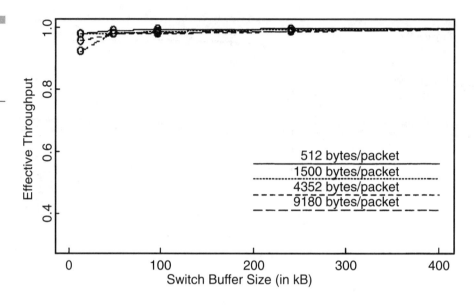

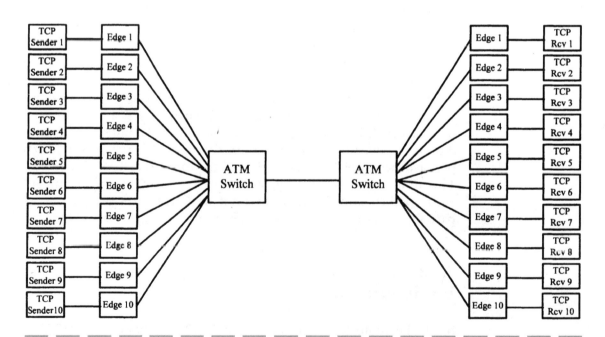

Figure 9-8. 10 TCP senders, 10 TCP receiver, no traffic concentration at edge devices.

connected by an OC-3 link. In this configuration the interswitch OC-3 is the bottleneck link.

This study compared the performance of UBR and ABR under the following conditions:

- UBR with Plain ATM
- UBR with EPD
- ABR: Explicit Rate ABR
- ABR with Front Drop (FD): In this scheme the edge device has the ability to drop both cells and packets from its output queue. When the edge device queue is full, FD drops a complete packet from the front of the queue (as opposed to the tail). The motive for doing this is that it allows earlier detection of congestion.

Round trip delay was 18ms on the TCP control loop and 6ms in the ABR loop. In [9] MTU size was 1,500 octets and the TCP window size was 640kB. All link rates (between TCP sources and ATM edge devices, and between ATM switches) were OC-3. Edge buffer size was 500 cells and 1,000 cells for ATM switches. The EPD threshold was 700 cells. Simulation results were based upon transmission of a 100kB file from sender to receiver. The simulation results present delay[3] and goodput[4] and are shown in Table 9-4.

Since each edge device serves only a single TCP source, congestion with UBR will be within the ATM switch serving the TCP sources. EPD improves both goodput and delay over the plain ATM case. ABR, however, pushes delay out of the network to the edge devices where edge buffers quickly fill up ultimately leading to discarded cells. Thus delay increases and goodput falls due to increased packet retransmissions. ABR with FD provides a significant improvement in both of these measures which

TABLE 9-4.

MTU = 1,500 octets

	UBR	UBR w/EPD	ABR	ABR w/FD
Delay	370 ms	230 ms	715 ms	237 ms
Goodput	63.4 Mbps	79.1 Mbps	36.15 Mbps	67.9 Mbps

[3] The time duration from transmission of the first octet of a 100KB file to receipt of an acknowledgment for the last octet.

[4] The rate of receipt of good packets (excluding incomplete and retransmitted packets) at the TCP layer.

suggests that it is not sufficient to simply move congestion out of the ATM network to improve TCP over ATM performance. The implication of this result is that an end-to-end congestion strategy that works in concert with ABR is required to realize the benefits from reduced ATM network congestion. Without such coordination, poor interaction between the TCP window and ABR flow control mechanisms results in significant TCP performance degradation. Even still, the results of Table 9-4 indicate that UBR with EPD provides the best TCP performance.

As suggested in [10], an alternative strategy would be to increase edge buffer memory. In [9] it was found that ABR with FD and a larger edge buffer provided better delay and goodput performance than UBR w/EPD. This does have the drawback that the amount of buffer required in an edge device to improve performance with a single TCP source must be increased proportionately when multiple TCP sources share a common edge device. Another drawback in large edge buffers is that they result in increased RTT as those buffers fill.

In [11] the focus was on the effect of varying buffer size on TCP performance under ABR and UBR. This study also extended [9] to include configurations with multiple TCP sources connected to a common edge device (Figure 9-9), and, in a separate simulation, introduced a third ATM

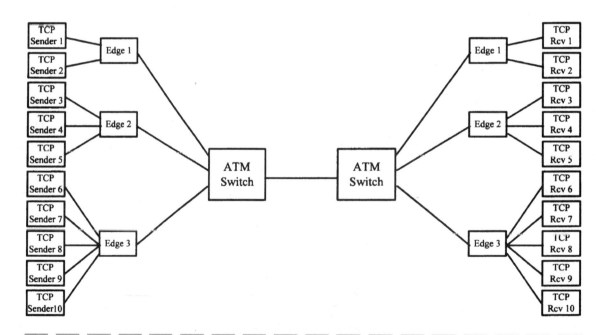

Figure 9-9. 10 TCP senders, 10 TCP receivers traffic concentration at edge devices.

Chapter 9: TCP OVER ATM

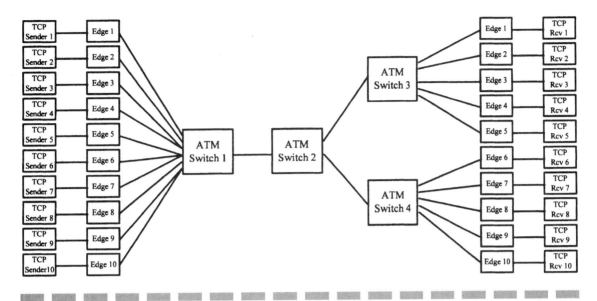

Figure 9-10. 10 TCP senders, 10 TCP receivers. 50 Mbps available bandwidth on switch 2 to switch 3 link.

switching stage (Figure 9-10). This study also used a 9,180 MTU size as in [10]. In the configuration of Figure 9-8 with ABR, a 700 cell edge buffer was sufficient to prevent edge buffer overflow during 100kB file transfers. In this case, ABR provided superior performance to UBR since ABR prevented cell loss within the ATM network, and the edge buffer was large enough to prevent packet loss. In UBR with EPD an ATM switch buffer size of 3,500 cells was required to avoid buffer overflow in the switch. As expected, for UBR with plain ATM there is no upper limit on the amount of ATM switch buffer required to avoid buffer overflow.

In the absence of buffer constraints within the ATM switch, however, we would expect UBR to provide the best performance because each TCP source would be able to transmit at line rate without concern for dropped cells. ABR, on the other hand, constrains the transmission rate from the edge devices to the current ACR.

In the configuration of Figure 9-9 the edge devices support up to 5 TCP sources. In this case, it was found that with ABR, an edge buffer size of up to 3,500 cells was required to avoid edge buffer overflow. For UBR, the required edge buffer size to avoid buffer overflow was 1,000 cells. The ATM switches are not required to support as many devices in this configuration. Thus, an ATM switch buffer of 2,000 cells was sufficient to avoid switch buffer overflow. Furthermore, with a switch buffer of 3,000 cells it was possible to realize the best performance with UBR and EPD if the

EPD threshold was appropriately chosen. This appropriate value was found to be 2,500 cells, 83% of capacity. TCP performance with UBR decreased with lower EPD thresholds. For EPD thresholds of 2,000 cells or less with a 3,000 cell switch buffer, TCP performance was better using UBR with plain ATM even though there was occasional switch buffer overflow in this case.

In the configuration of Figure 9-10, a third level of switching is introduced. Furthermore, background traffic on the switch 2 to switch 3 link reduced available bandwidth to 50Mbps. In this configuration, the bottleneck link for TCP sources 1—5 was the link between switch 2 to switch 3. For TCP sources 6—10 the bottleneck link was the link between switch 1 to switch 2.

The simulation set a large buffer capacity at switch 1 for the UBR case to observe cell dropping behavior at switch 2. In this case, overall TCP performance suffered because bandwidth was occupied in the switch 1 to switch 2 link by cells which were ultimately dropped due to congestion in the switch 2 to switch 3 link.

ABR limited bandwidth utilization by TCP sources 1—5 to about 1/3 of the bandwidth on the switch 1 to switch 2 link.

In the Figure 9-10 configuration, connections that were routed over the switch 2 to switch 3 link had higher throughput with ABR than with UBR. Conversely, connections that were routed over the switch 2 to switch 4 link had the full OC-3 bandwidth available over this link. Thus, these connections performed better with UBR than ABR.

4. Conclusion

ABR and TCP use independent flow control mechanisms, and additional delays are introduced in allowing the TCP source to react to congestion within the network. There is a delay within the ABR control loop, followed by a second delay in the TCP control loop.

One proposed solution is to link these mechanisms at the edge device [12]. The edge device is to use information in the resource management cells to determine whether congestion exists within the network. In response, the edge device would delay forwarding of acknowledgment packets, which would trigger time-outs at the TCP source. This proposal does require the edge device to keep track of (or otherwise have knowledge of) the retransmit timers in each TCP source. This proposal does, in theory, provide faster congestion notification to the TCP source since it is coming directly from the congested node (through the RM cell).

Another possibility is development of TCP over native ATM implementations. Such TCP implementations would be able to use information from both TCP control and ABR loops. This proposal, however, would require development of TCP implementations, which are based upon the underlying data link layer—a violation of basic OSI layering principles.

TCP could be tuned for ATM network implementation to provide faster congestion detection and response. For example, in connectionless packet networks each packet can take a different route through the network producing (potentially) out-of-sequence packet delivery. Thus, a threshold of three duplicate acknowledgments is used to allow some margin for the possibility that later sent packets arrived sooner than one sent earlier. ATM, however, is a connection-oriented technology that guarantees in-sequence delivery. Consequently, the first duplicate acknowledgment could trigger a fast restart. This faster response could, in turn, further reduce the likelihood of time-out and slow start, and potentially improve TCP performance over ATM. This would, however, be a TCP implementation that would be dependent upon a particular technology in the underlying network.

In summary, it may be suggested that for strictly TCP/IP-oriented data applications, ATM provides no inherent performance benefits over packet networks. ATM does introduce additional complexity: large required switch buffers for UBR and the need for coordination between TCP and ATM flow control mechanisms when ABR is used.

References

1. Lindberg, Gunnar, and Robert Olsson, Report from COAST Phase 1, Assorted Experiments with TCP/IP over ATM Wide Area Links, Swedish University Network (SUNET), September 1995.
2. Comer, Douglas, *Internetworking with TCP/IP 2d Ed.*, Prentice-Hall, 1991.
3. England, Kent, ATM Performance Requirements for TCP, October 1996.
4. Postel, Jon, Transmission Control Protocol—DARPA Internet Program Protocol Specification, Internet Engineering Task Force, RFC 793, September 1981.
5. Jacobson, Van, Bob Braden, and Dave Borman, TCP Extensions for High Performance, Internet Engineering Task Force, RFC 1323, May 1992.

6. Mogul, Jeffrey, and Steve Deering, Path MTU Discovery, Internet Engineering Task Force, RFC 1191, November 1990.

7. Jacobson, Van, Congestion Avoidance and Control, In *Proceeding of SIGCOMM '88*, ACM.

8. Romanow, Allyn, and Sally Floyd, Dynamics of TCP Traffic over ATM Networks, IEEE Journal on Selected Areas in Communications, May 1995.

9. Jagannath, Shantigram, and Nanying Yin, More Results on End-to-End Traffic Management in IP/ATM Internetworks, ATM Forum, Document 97-0653, July 1997.

10. Jagannath, Shantigram, and Nanying Yin, End-to-End TCP Performance in IP/ATM Internetworks, ATM Forum, Document 96-1711, December 1996.

11. Ren, Wenge, Kai-Yeung Siu, Hiroshi Suzuki, and Gopalakrishnan Ramamurthy, Performance of TCP over ATM with Legacy LAN, ATM Forum, Document 96-1716, December 1996.

12. Ma, Jian, Interworking Between TCP and ATM Flow Controls, ATM Forum, Document 97-0960, December 1997.

CHAPTER 10
ATM and the Internet

1. Introduction

By mid-1997 the Internet included more than 1.3 million networks which interconnected over 26 million computers which served an estimated 101 million users throughout the world [18]. Rapid growth of traffic has sparked concerns about the possibility of congestion collapse, a condition in which packet retransmissions flood an already congested network driving effective throughput to virtually zero, on the Internet.

By 1998, a reported 67 percent of Internet traffic in the United States was switched over an ATM network at some point in transit [1].

In this chapter we will discuss how ATM technology is being deployed in the Internet.

2. Evolution of the Internet

The Internet has its origins in the late 1960s with the ARPANET. During the 1970s the network grew to support many organizations in the U.S. Department of Defense, other government agencies, and universities and research organizations [2]. The development of TCP/IP in the early 1980s marked the beginning of the Internet.

Starting in 1985, the National Science Foundation (NSF) began to take an active role in expanding the Internet among the research community. The NSF initiated a new long-haul backbone network, called the NSFNET, which was initially based upon 56-kbps links that connected 6 NSF supercomputer centers and several university-based regional networks. The geographic reach of this backbone network spanned the continental United States.

Within months after its inception, the NSFNET backbone was overloaded. In 1987 an award was granted to a partnership consisting of Merit Network, Inc., IBM Corporation, and MCI, Inc. This new NSFNET backbone was based upon T-1 circuits, and connected 13 regional networks. In 1990, Merit, IBM, and MCI spun off a new organization known as Advanced Network Services (ANS), which had the charter to commercialize and upgrade the backbone network, providing NSFNET service along with other customer traffic on the same network. Merit Network's Internet Engineering group provided a Policy Routing Database and routing consultation and management services for the NSFNET, while ANS operated the backbone routers and a Network

Operations Center. By 1993, ANS and Merit performed a major upgrade of the backbone to DS-3 circuits.

Beginning in the late 1980s, multiple U.S. government agencies and commercial organizations became network service providers, creating national backbone networks for various purposes. The NSFNET backbone service was intended for research and educational applications only. By 1992, the bulk of U.S. educational and research organizations were connected to the NSFNET. But the amount of traffic and number of organizations utilizing the NSFNET were still growing. Government agencies had interconnected at Federal Internet Exchange points on the east (FIX-East) and west (FIX-West) coasts. Commercial network organizations had formed the Commercial Internet Exchange association, which built an interconnection point on the west coast (CIX). Internet providers in Europe and Asia and on other continents had also developed substantial infrastructures and connectivity. In 1992, the NFS selected Sprint as NSFNET international connections manager providing connectivity between the NSFNET and international research and education networks. Other U.S. networks also interconnected with international networks.

In 1994, NFS awarded contracts to replace the NSFNET with a new Internet backbone architecture—one not operated by the NFS. This architecture consisted of:

- Network Service Provider (NSP). An NSP maintains a backbone network providing broad Internet connectivity nationwide. Instead of a single Internet backbone operator, the new Internet architecture is based upon competing backbone service providers.

- Very high performance Backbone Network Service (vBNS). vBNS maintains a high-speed Internet backbone network that supports research, scientific, and educational purposes. vBNS also serves as an initial deployment site for leading edge network technologies.

- Network Access Point (NAP). There are points of interconnection for NSP networks and regional Internet Service Providers (ISP). These NAPs allow the exchange of Internet traffic among NSPs and ISPs. Such exchanges are referred to as *peering*. The NSF sponsored four NAPs (located in New York City, Washington D.C., Chicago, and San Francisco), however, due to the explosive growth of the Internet, additional unsponsored NAPs, CIXs, and private exchange points have also appeared.

- Routing Arbiter (RA). The RA provides address resolution and routing databases for the exchange of Internet traffic among NAP-attached ISPs. The NSF maintains a RA at each sponsored NAP. An operations center assists in maintaining operational stability in the Internet, similar to the function provided by Merit under the NSFNET.

The NSFNET backbone was replaced by the new architecture in 1995.

3. Internet Standards

The principal body engaged in the development of Internet standard specifications is the Internet Engineering Task Force (IETF) [3]. The IETF began in 1986 and is made up of volunteers who meet to fulfill the mission and objectives of the organization. There are two types of documents issued by the IETF. The IETF operates under the auspices of the Internet Society (ISOC) board of trustees which appoints members of the Internet Architecture Board (IAB) to provide oversight for the Internet standards process.

4. Internet Traffic Patterns

A May 1992 study of NSFNET traffic patterns reported that the dominant Internet applications were FTP, SMTP, NNTP, DNS, and Telnet, along with a growing amount of "other" traffic [4]. Packet sizes indicated a mixture of bulk data transfers and interactive applications with a mean packet size of 186 octets.

By 1995 traffic patterns were starting to change. Information storage and retrieval applications, such as the World Wide Web were beginning to overtake electronic mail and file transfers in network traffic volume [5]. As shown in Figure 10-1, the leading identified application by early 1995 was WWW at 21% of NSFNET backbone traffic.

A 1997 study of Internet traffic over an NSP backbone indicated that the WWW has increasingly dominated network traffic [6]. In this study, traffic at two measuring points was monitored:

Chapter 10: ATM and the Internet

Figure 10-1.
Internet traffic share by application (1995).

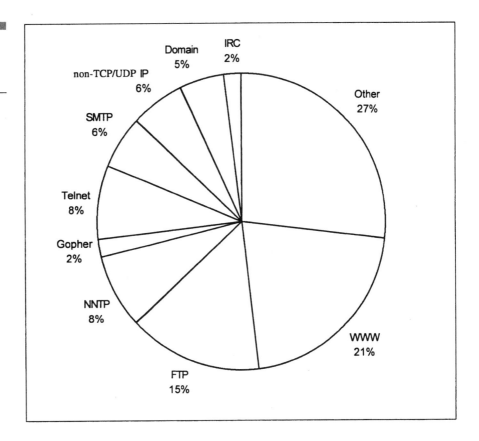

1. Domestic traffic was measured at an NAP near a major U.S. East Coast city.
2. International traffic was measured at an interconnection point that served a transatlantic link.

At the domestic measuring point, traffic volumes varied from just over 10 Mbps to nearly 55 Mbps according to the time of day. On the international link traffic volumes varied between just under 5 Mbps to over 40 Mbps.

At the domestic measuring point, the Web accounted for 75% of octets and 70% of packets carried over the NSP backbone. TCP was the dominant protocol accounting for 95% of octets and 90% of packets. 75% of packets were smaller than 552 octets with half of all packets between 40 octets and 44 octets in length. While packet sizes range to as high as almost 4,500 octets, nearly all packets were 1,500 octets or smaller.

5. The Integrated Services Internet

As originally conceived, the Internet offered only a single QoS: "best effort" delivery. Thus, packets from all sources were treated equally with resources allocated to traffic on a first-come first-served basis [7]. However, best-effort delivery does not support real-time applications, which require the ability to control end-to-end delay.

In addition, demand for a modified Internet infrastructure arises from network operators who wish to control the sharing of bandwidth on a particular link through the ability to classify traffic into different traffic categories. Thus, each traffic category could be allocated a determined amount of minimum bandwidth under network congestion while allowing additional link bandwidth to be used on an as-available basis. Traffic categories may be based upon upper layer protocol (TCP, UDP, IPX, etc.), TCP/UDP port number (WWW, FTP, Telnet, etc.), or upon user group. Such a management facility is referred to as *controlled link sharing*. The Integrated Internet Service (IIS) model is a modification of the original Internet service model that supports best-effort service, real-time service, and controlled link sharing [8].

The multiple QoS supported by the IIS is based upon the fundamental concept of identifying individual packets as belonging to a *flow*, which refers to a stream of packets that result from a single end-user activity (such as an application) that requires the same QoS. Since a flow is identified at the source, there may be a single destination or multiple destinations (as in a multicast). Unlike ATM, IP is a connectionless protocol so the closest analogy to an IP flow in the ATM protocol is the ATM VC.

Routers that are to implement the IIS must perform traffic control to implement the appropriate QoS for each flow. Traffic control consists of the following basic functions, as shown in Figure 10-2:

- Packet Scheduler
- Packet Classification
- Admission Control
- Resource Reservation
- Routing

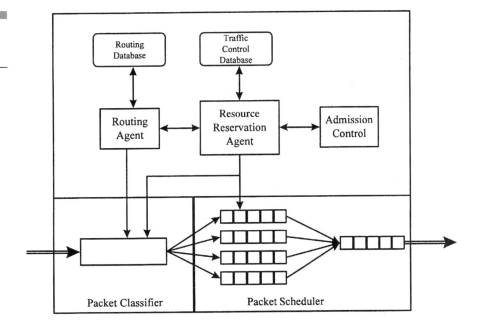

Figure 10-2.
Internet integrated services router.

5.1 Packet Scheduler

The packet scheduler is responsible for managing the forwarding of different packet streams to determine the timing and order of packet forwarding for each flow being served. Packet scheduling is implemented at the output ports using a set of queues to which packets from different traffic categories are directed. The packet scheduler then determines the order in which packets from the various queues are dispatched.

A simple approach to packet scheduling is to assign fixed priorities to each queue. Thus, packets in the highest priority queue have absolute priority over packets in lower priority queues regardless of when the packet arrived in the queue. It is possible in this scheduling regime for high priority queue arrivals to completely block lower priority packets from being forwarded.

An alternative scheme is to use Weighted Fair Queuing (WFQ) which assigns to each queue a percentage of link bandwidth. Under WFQ the packet scheduler services each queue in a round robin fashion forwarding any waiting packets until the allotted link bandwidth allocation has been exhausted. If a queue has no pending packets to forward the available bandwidth may be used to forward packets from queues that do.

In Figure 10-3 we see an example of WFQ where Queue_A gets a 60% allocation of link bandwidth, Queue_B gets 30%, and Queue_C gets 10%. So long as there are packets in all queues, each queue receives its allotted bandwidth on the link. When Queue_A is emptied, the remaining queues receive proportionate shares of the available, but unused bandwidth.

Figure 10-3.
WFQ example.

Queueing Schedule

Time slot	1	2	3	4	5	6	7	8	9	10
Queue	A	A	A	A	A	A	B	B	B	C

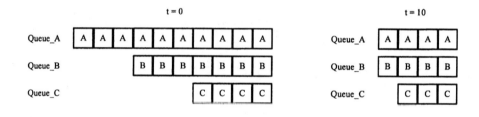

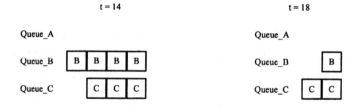

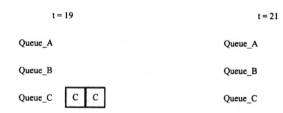

When served, each queue still gets the same bandwidth allotment as before on each pass through the round robin scheduler. However, the packet scheduler returns to serve the remaining queues more frequently since there are no packets pending forwarding in Queue_A.

It is worth noting that WFQ imposes a delay upon packet forwarding time, which is a function of the queue servicing allocation:

$$q_d = \frac{t_d \times q_x}{wfq_x}$$

where q_d represents the maximum queuing delay, t_p is the transmission time for forwarding of a single packet, *wfqx* is the bandwidth allocation for *Queue_x*, and q_x is the length of the packet queue.

In ATM the analog to the IIS packet scheduler function is traffic shaping.

5.1.1 Packet Discard

Related to packet scheduling, in the achievement of a desired QoS, is the issue of packet discard as a router must drop packets when its buffers are full. As discussed in the last chapter, a dropped packet is detected by TCP as a time-out while waiting for an acknowledgment. This is interpreted as an indication of congestion, which throttles the source. Therefore, packet dropping strategy relates to selection of which source is to be throttled.

Within the context of real-time services, however, the objective of packet discard is to provide the desired QoS for each flow. In queues intended to support real-time services an upper limit may be enforced on queue length. Packets that arrive after the queue length threshold has been exceeded are subject to discard[1]. At a simple level, limiting queue length achieves the objective of limiting packet forwarding delay since the longer an output queue is allowed to become, the longer the arriving packet will have to wait before actually being forwarded on the output link. In reality, the delay experienced by a packet is also dependent upon the packet scheduling strategy. As suggested above, a packet scheduling algorithm based upon hard queue priority assignments could produce indefinitely long delays in lower priority queues independent of

---

[1] This is referred to as "tail drop" since the most recently arriving packet (i.e. the one at the tail of the queue) is dropped.

the length of the queue. Thus, we see that packet discard and packet scheduling strategies must be coordinated.

A simple tail drop strategy has two significant disadvantages:

1. Depending upon the synchronization between flows from different sources, there are cases where a single flow, or group of flows will be able to monopolize queue space.

2. Tail drop strategies allow queues to be maintained in a full, or near full, state for extended periods of time. Setting queue thresholds such that queues are maintained in a nonfull state has the benefit of reducing end-to-end delay at the expense of making the network less able to absorb bursts of packet traffic. This can result in significant degradation in throughput as an arriving burst may cause multiple packets to be dropped from many, or all, flows sharing the queue. Since multiple packet drops can trigger TCP slow starts, the net effect would be to cause all flows to throttle back at the same time. In the worst case, flows become synchronized which would produce a cyclical congestion-slow start pattern. Thus, a delay versus throughput trade-off in this instance introduces a bias against lower average bandwidth, but highly bursty flows.

One packet dropping algorithm which addresses these shortcomings is Random Early Drop (RED) [9][10]. The RED algorithm assigns a probability of drop to packets arriving in a queue. To first order, the probability increases as the average length of the queue increases. The RED algorithm is typically implemented to estimate the average queue size based upon an exponentially weighted moving average:

$$avg_{t+1} = (1 - w_q) \times avg_t + w_q \times q_t$$

where avg_t is the weighted average queue length at time t, q_t is the current queue length, and wq is the weighting factor. The weighting factor determines the responsiveness of the average queue size estimate to packet bursts. By selecting a sufficiently small weighting factor, the average queue size will not be very sensitive to short packet bursts. However, weighting factors that are too small make the estimate too slow to respond to sustained changes in packet traffic. This could cause queues to fill before the algorithm is able to detect it, which renders the algorithm ineffective in queue management.

The other part of the RED algorithm is in the determination of which packets to drop. The RED algorithm uses a minimum queue threshold parameter, *minth*, such that all packets are forwarded for average queue length less than *minth*. When the average queue length exceeds a maximum queue threshold, *maxth*, all packets are dropped. For average queue lengths between these bounds, a probability of packet drop is assigned to each packet, which increases with average queue length. A simple model would have this probability, p_1, increase linearly up to a maximum probability, *maxp*, for average queue length between *minth* and *maxth*:

$$p_1 = maxp \times (avg - minth)/(maxth - minth)$$

The probability of packet drop as a function of average queue length is shown in Figure 10-4 for maxp = 10%, minth = 5, and maxth = 12.

A variation of the RED algorithm is referred to as Weighted Random Early Drop (WRED), which assigns packet drop probabilities that differ on a per-flow, or a per-queue basis. This allows high priority flows to have less chance of being dropped relative to lower priority flows.

In comparison to ATM, WRED is analogous to Usage/Network Parameter Control.

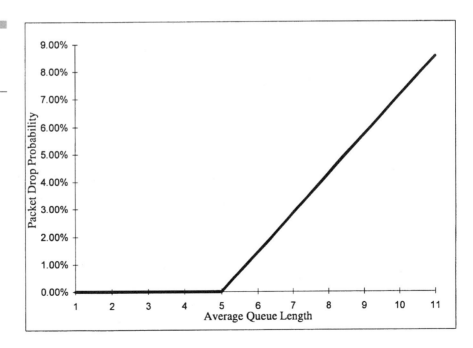

Figure 10-4.
Random early drop: packet drop probabilities.

5.2 Packet Classification

The objective of packet classification is to assign each incoming packet to a class for treatment by the packet scheduler. All packets in a given class receive the same treatment from the packet scheduler. Packet classification is local to the individual router only; the same packet may be classified differently by each router along an end-to-end route. For example, backbone routers in the core of a network may map many flows into a relatively small number of aggregated classes while routers nearer to the traffic sources may use a separate class for each flow. Packet classification may also influence the routing of the packet.

There are a number of methods that may be employed in implementing packet classification. At the very minimum, packet classification requires that the router look at more than the destination IP address. For example, the packet classifier could use Layer 3 information in the IP header such as source/destination IP address, and protocol identifier (TCP, UDP, etc.). Layer 4 information could also be used, such as the TCP/UDP port address, which identifies the type of application (WWW, FTP, SMTP, or NNTP).

IPv6 provides a "flow-id" field, which reduces the overhead associated with packet classification from that described above [11]. The flow-id can be used to act as a handle that could be cached in the router to provide relatively quick packet classification.

In the ATM-equivalent to packet classification in IP networks, cells are "classified" based upon VPI/VCI header information.

5.3 Admission Control

Admission control is the set of functions that determine whether a new flow may be granted a requested QoS without negatively impacting QoS guarantees granted to existing flows. Admission control is performed by routers in response to a request from a traffic source that network resources be reserved for a specific flow.

Admission control in IP networks is similar to connection admission control in ATM networks with the critical difference that IP networks do not establish connections. Thus, even if a reservation request is refused, the source is still able to send packets through the networks, albeit without QoS guarantees on the flow.

Closely related to admission control (and often considered to be a part of it) is policy control, which is concerned with enforcing administrative policies on resource reservation requests. These administrative policies are concerned with issues such as authenticating the user making the request and verifying that they have permissions to make the resource reservation request.

One method that has been proposed for the establishment of resource reservations within IP networks is the Resource Reservation Protocol (RSVP).

6. Resource Reservation Protocol

RSVP is used by routers to deliver QoS requests to all nodes along the path of flows and to establish and maintain state information to provide the requested service. RSVP is a transport layer protocol that operates with either IPv4 or IPv6 [12]. Unlike transport protocols such as TCP or UDP, RSVP does not transport application data but instead acts as an Internet control protocol similar to the Internet Control Message Protocol (ICMP), or the Internet Group Management Protocol (IGMP).

RSVP is not a routing protocol but is designed to work with both point-to-point and multicast routing protocols. RSVP consults routing databases to determine how to route resource reservation requests along the flow path. In the multicast case, a host would send IGMP messages to join a multicast group and also send RSVP messages to reserve resources along a delivery path for that group. Thus, routing protocols still determine where packets get forwarded; RSVP sits in the background maintaining state information about the resources that are currently being used along the routes over which packets are being forwarded.

RSVP acts as a middleman process, receiving RSVP messages and querying other processes which process the information in the RSVP messages and report status information back to RSVP. In RSVP resource reservation requests are sent, hop by hop, from the receiver upstream to the traffic source along the reverse route of the preceding RSVP Path message. Resource reservation decisions are made at each hop independently from downstream routes until the RSVP Resv message reaches the traffic source or is rejected by an intermediate router along the path. At each router the admission control process determines if the resource reservation is to be accepted. In the local router node, RSVP also sends information to both the packet scheduler and packet classifier processes to set up the appropriate QoS for handling of packets associated with the flow within the router.

RSVP is a simplex protocol in that RSVP messages apply in only one direction. Thus, there are RSVP sender messages and RSVP receiver messages:

- Senders use RSVP Path messages to advertise a route to the receiver and to provide information about the resources that are available along the route. The Path message contains information that describes the data traffic generated by the sender (the Sender TSpec). The Path message also carries information generated within the network that describes available services, delay and bandwidth estimates, and other operating parameters that are specific particular QoS services. This information is referred to as an ADSPEC.

- Receivers use RSVP Resv messages to reserve resources to receive information from the sender with a requested QoS. The receiver sends information in the Resv message to describe the traffic flow to which the resource reservation should apply (the Filter Spec), and a description of the desired QoS (the Flowspec). Filter spec information is used by the packet classifier while Flowspec information is used by the packet scheduler. The Filter Spec and Flowspec form a Flow Descriptor.

There are a number of critical differences between the method of QoS assignment under RSVP from that of ATM:

1. ATM is connection oriented, so once QoS is granted for the connection the state is maintained for the duration of the connection. In RSVP, there are no connections, instead "soft" state is maintained in which receivers are granted resource reservations on selected flows for limited periods of time only, after which the resources are released. Receivers wishing to maintain resource reservations must send additional Resv messages within the allotted reservation lifetime to extend or modify resource reservation parameters.

2. In ATM, either the sender or receiver may initiate a connection at which point QoS is assigned. In RSVP, only the receiver may make a QoS request.

3. In ATM, if a connection with the requested QoS is rejected by either the network or the receiver (during connection establishment) there can be no traffic flow between the sender and receiver. In RSVP if a

resource request is rejected, the flow receives "best effort" QoS by default. The receiver may reattempt the resource request until it is granted at which point the flow receives the requested QoS.

4. ATM traffic management enforces compliance with the traffic contract such that noncompliant traffic is discarded. In RSVP, traffic that exceeds the resource reservation receives "best effort" delivery.

5. In multicast communications, Flowspecs from multiple receivers are merged at routers along intermediate hops to produce a composite Flowspec to the sender which contains a set of resource reservations that are sufficient to accommodate the resource reservations at each of the downstream receivers. Thus, individual leaf nodes that are members of the same multicast group may receive different QoS. In ATM multicast, each leaf node must receive the same QoS.

At branch points in a multicast distribution, it is possible that the Flowspecs for the various branches will not be the same. This is called a *heterogeneous branch point*. It may be necessary to perform traffic reshaping to ensure that all outgoing traffic from the heterogeneous branch point (which is also referred to as a reshaping point) is in compliance with the respective Flowspecs of each multicast branch.

It should be added that while RSVP operates over the connectionless IP protocol, it does give the appearance of a connection. Path messages contain information which allows subsequent Resv messages to set up a resource reservation request along the route defined by the Path message.

6.1 Path Message

RSVP supports a number of message types. One such message is the Path message, which is transmitted by a sender of traffic into the network. The contents of the Path message, and all other RSVP messages, is a set of objects that contains information that is appropriate to the message. Key objects in the Path message are:

- Session: the Session field identifies the destination IP address, Protocol, and Transport Layer Destination Port to which the traffic source is to send packets associated with this flow.

- RSVP_Hop: contains the IP address of the last RSVP hop that forwarded the Path message. A node receiving this message caches this information. Thus, the receiving endpoint of the flow route has a means of backtracking (on a hop by hop basis) the route to the traffic source.
- Sender Template: contains the IP address and port for the traffic source.
- Sender Tspec: defines the traffic characteristics of the traffic flow.
- Adspec: contains advertisements of network service offerings and network performance.

Details on the contents of the Sender Tspec and Adspec are discussed in the following sections.

6.1.1 Sender Tspec

The sender Tspec carries information about the data source's generated traffic. It consists of:

- r: token bucket rate (octets/second)
- b: token bucket size (octets)
- p: peak data rate (octets/second)
- m: minimum policed unit (octets)
- M: maximum packet size (octets)

Sender Tspec parameters inform the receivers of the characteristics of the traffic transmitted on the flow. The parameter, r, represents the normal rate at which the source, or any reshaping point, can inject traffic into the network. The parameter, p, indicates the peak rate at which the source or reshaping point may inject traffic into the network in bursts. The duration of such bursts is limited by the token bucket size parameter, b. The parameters, r, b, and p are analogous to the SCR, MBS, and PCR parameters from ATM. The m parameter represents the minimum packet size recognized in the flow. Thus, for policing packets smaller than m are treated as though they were of size m. M represents the maximum MTU size allowed in the flow. Segmentation of packets is not allowed, thus, any packet larger than M is considered noncompliant and therefore not eligible to receive the QoS associated with the flow.

6.1.2 Adspec

The Adspec carries two types of information:

1. General characterization parameters pertaining to the network, independent of QoS:
   - Available Path bandwidth along the route followed by the flow. While evaluated at each router on a hop-by-hop basis, the end-to-end path bandwidth estimate is the minimum estimate from among the router nodes.
   - Minimum path latency—this delay includes factors such as propagation delay but does not include queuing delays. Each router evaluates minimum latency through the node with the end-to-end minimum path latency representing the cumulative value of the minimum latencies of the individual nodes in the path.
   - Path MTU, which is the maximum transmission unit for packets following the route of the flow. This value is necessary for QoS services that require that the MTU be limited to a specific size. Each router assesses its MTU value with the end-to-end Path MTU being the minimum MTU among all routers in the path.

2. Service specific information that is related to QoS. There are two types of IIS services [13][14]:
   - Guaranteed Service is a service that offers bounded end-to-end delay to traffic that is in compliance with the Resv message. Guaranteed service parameters include:
     - Ctot: this represents an end-to-end measure of nonlatency related delay through each router which is related to the rate of traffic flow. An example would be the time it takes to serialize a packet for transmission. This parameter is expressed in octets, thus, the corresponding time delay is measured by Ctot divided by the relevant transmission rate.
     - Dtot: this represents an end-to-end measure of nonlatency related delay which is not directly related to the transmission rate. An

example would be delay in packet forwarding due to a queuing scheme such as WFQ. This parameter is expressed in microseconds.

- Csum: the equivalent of Ctot but measured between contiguous reshaping points only.

- Dsum: the equivalent of Dtot as measured between contiguous reshaping points.

Controlled Load Service is a service that does not offer bounded end-to-end delay to traffic that is in compliance with the Resv message. Controlled Load service does provide a higher priority delivery than Best Effort in that compliant traffic receives an allocation of reserved network resources.

The relationship between Guaranteed Service parameters is shown in Figure 10-5. Ctot and Dtot allow receivers to determine the maximum end-to-end queuing delay. Csum and Dsum are used at reshaping points to determine the amount of buffering required to avoid packet loss.

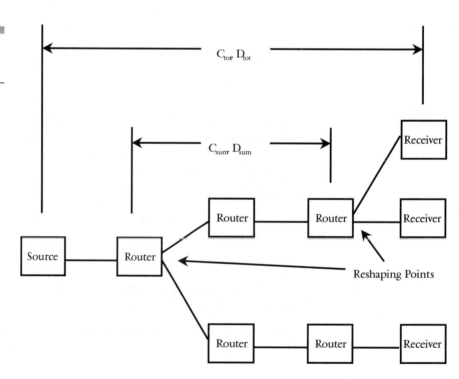

Figure 10-5.
Delay parameters in RSVP path.

6.2 Resv Message

Key objects in the Resv message are:

- Session: in the Resv message, the destination IP address is the RSVP_Hop address received in the corresponding Path message.
- RSVP_Hop: contains the IP address of the last RSVP hop that forwarded the Resv message. This object creates a trail to the receiver over which a reservation confirmation message may be sent to the originator of the Resv message.
- Filter Spec: contains the IP address, and port for the traffic source.
- Flowspec: contains the information necessary to make the reservation request.

6.2.1 Flowspec

The Flowspec consists of two sets of parameters: a Rspec (R for "Reserve") that describes the desired QoS, and a Tspec (T for "Traffic") that describes the data flow. The Receiver Tspec carries information that reflects the receiver's desired reservation. While the Flowspec differs between the Guaranteed Service and Controlled Load Services the following parameters are common:

- r: token bucket rate (octets/second)
- b: token bucket size (octets)
- p: peak data rate (octets/second)
- m: minimum policed unit (octets)
- M: maximum packet size (octets)

These parameters have the same interpretation as in the Sender Tspec. In the Resv message, however, these parameters are evaluated at each hop as a request for resource allocation rather than a traffic description as is the case with the Sender Tspec. In the Flowspec, the value M should be set to a value no greater than the Path MTU value received in the preceding Path message. The token bucket rate in the Receiver Tspec should be no less than the corresponding value in the Sender Tspec. However, should this not be the case, surplus sender traffic above the reserved rate is subject to random packet loss. The peak data rate in the Receiver Tspec should be less than the Available Path Bandwidth value advertised in the Adspec [15].

In addition to the Receiver Tspec, the Guaranteed Service specifies a Receiver Rspec, which has the following parameters:

- R: a rate greater than or equal to the r parameter from the Tspec
- S: a slack term (microseconds)

The difference between the token bucket rate from the Receiver Tspec and the R parameter from the Rspec is one of perspective. The token bucket rate is a reservation for network resources that allows a source, or reshaping point to inject traffic at the token bucket rate with a reserved amount of buffering at the source being determined by the token bucket size parameter of the Receiver Rspec. The reality is that there is also buffering required at intermediate hops along the route between the sender and receiver. Thus, the motivation for the Rspec is to achieve a specific end-to-end delay bound. The token bucket, b, and error terms Ctot and Dtot suggest a delay of:

$$b/R + C_{tot}/R + D_{tot}$$

R can be set to a value to achieve the desired end-to-end delay by reserving resources for a sufficiently high data rate, which offsets the delays encountered due to buffering within router hops. If the error terms were not present (i.e., no delays within the network, save the inevitable latency/speed of light delay), delay could be characterized at the source as being equal to b/r. Thus, to conclude our point, the token bucket rate in the Receiver Tspec is established from the perspective of the traffic source, while the R parameter in the Receiver Rspec is established from the end-to-end perspective. The parameter R must be no less than the token bucket rate, and should be no greater than the Available Path Bandwidth.

The R parameter alone suggests that the reservation be made for an end-to-end flow with the corresponding data rate. In some cases, however, congestion may exist on some hops along the route such that a reservation requesting R will fail. The slack term, S, allows local nodes to vary the data rate along the route. Figure 10-6 shows that as the Resv message traverses the route from receiver to sender, a particular node may alter R_{in} in a receiver Rspec and produce a lower value R_{out}, which is then inserted in the Rspec as it moves upstream to the sender. As this also uses up some of the slack S_{in}, the outgoing Rspec also has a lower slack, S_{out}. In essence, the node consumes some of the slack according to the relation:

$$S_{out} + b/R_{out} + C_{tot}(i)/R_{out} \leq S_{in} + b/R_{in} + C_{tot}(i)/R_{in}$$

where $C_{tot}(i)$ represents the accumulated C_{tot} up to router node i.

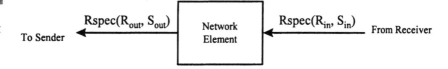

Figure 10-6.
Absorption of slack at a network element. Network element absorbs $(S_{in} - S_{out})$ amount of slack.

6.3 Initiation of an RSVP Session

The following is a possible series of steps that may be followed to create an RSVP multicast session:

1. A receiver joins the multicast group specified by an IP Class D Destination Address using IGMP.
2. A potential sender starts sending RSVP Path messages to the Destination Address.
3. A receiver application receives a Path message.
4. A receiver starts sending Resv messages to request resource reservations for receipt of the sender flow.
5. A sender application receives a Resv message.
6. A sender starts sending data packets.

7. Integrated Internet Services Over ATM

In this scenario an ATM network is in the end-to-end path of a network consisting of host, routers, edge devices, and ATM switches as shown in Figure 10-7. The edge devices are responsible for setting up and tearing down ATM connections by signalling as needed to support IP flows. The edge device must also perform VC management to determine if a flow requires a new VC or can use an existing one. If RSVP is used, the edge devices are to process RSVP messages and perform the appropriate translations into ATM signalling messages. Since RSVP is a receiver-oriented protocol that supports multicasting, this would suggest the use of ATM UNI 4.0 which supports leaf initiated join capability. This is not really the case, however, as the sender (or edge device on

Figure 10-7.
ATM network in RSVP path.

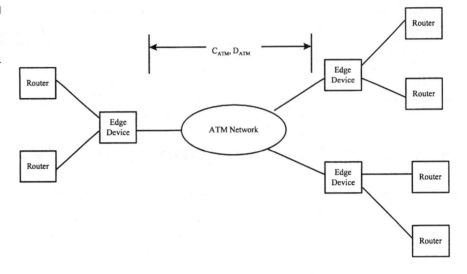

the upstream side of the ATM network) can initiate an SVC upon receipt of an Resv message.

In this section we will discuss the two major issues regarding IIS and ATM interworking: VC management and QoS management.

7.1 VC Management

7.1.1 PVC versus SVC

The primary objectives of VC management are to determine how many VCs are needed to support IIS flows and to map each of those flows to the appropriate ATM VC [16]. Either PVCs or SVCs may be used. If PVCs are used, the ATM network may be modeled as consisting of a series of point-to-point links, similar to the topology of many packet networks. In a PVC based network, the number of PVCs required to support IIS can become prohibitive for large networks of edge devices: $N(N-1)/2$ PVCs are required to provide fully meshed connectivity among N edge devices.

SVCs may be used to allow VCs to be established on demand, but if a single SVC is established for each flow, the volume of signalling traffic from RSVP resource reservations and refresh messages poses potential scaling problems within ATM networks, which may not be able to support the rate at which signalling messages are generated, or may have limits upon the total number of VCs that may be supported concurrently.

7.1.2 ATM Multipoint Connections and RSVP Heterogeneous Branch Points

ATM point-to-multipoint VCs may support RSVP point-to-multipoint distribution to receivers if all receivers have the same reservation at the branch point (e.g., a homogeneous branch point). However, RSVP supports the notion of heterogeneous branch points where each multipoint leg may have a different QoS. Support for this feature requires multiple VCs within the ATM network.

Potential options for support of multicast flows are:

1. Point-to-point VCs between the source and each receiver
2. Point-to-multipoint VCs between the source and each set of receivers with the same QoS
3. Single Point-to-Multipoint VC for each Multicast Group

7.1.2.1 Point-to-Point VC Point-to-point VCs from the source to each receiver has the obvious advantage that each receiver is able to establish resource reservations independently from other receivers sharing the flow. The disadvantage within the ATM network is one of scalability for the number of VCs that must be supported over all multicast flows. Each sending-side edge device establishes a separate VC to each receiving edge device. This also means that the edge device must replicate packet traffic for each such VC in the multicast.

There is a range of options available in this approach, from the sender establishing a separate VC to each receiver for each IIS (guaranteed service, controlled load service, and best effort service), to the sender establishing a single VC to each receiver over which would be multiplexed all IIS QoS.

In IIS, a flow may make a reservation request in an RSVP Resv message; however, if the reservation is rejected, the flow is to receive best effort service. In the two options mentioned above it is possible that the ATM network may not be able to establish a VC to support a given flow. In this case, the receiver would be denied service, a result that is inconsistent with IIS. When separate VCs are established for each flow between sender and receiver, the VC could be rejected either because a fundamental limit has been reached (such as the bandwidth limit of the UNI), or because the number of VCs supported by a given ATM network element has been exceeded.

When the edge device multiplexes flows over a common VC, the edge device must limit resource reservations according to the resources available on the VC. Furthermore, as resource reservations fully utilize the resources on the VC, "best effort" service becomes "no service." Thus the receiver (jointly with the sender) must establish additional VCs as existing ones become saturated.

There is still, however, the potential for denial of service on best effort flows for extended periods of time if the resource reservations approach, but do not exceed, the resources available on the VC. This would suggest that a better model in this case would be to segregate best effort flows from guaranteed and controlled load flows on separate VCs between the sender and each receiver.

7.1.2.2 Point-to-Multipoint VC for Each QoS This approach produces a number of VCs that is never greater than in the point-to-point case and is only equal in the degenerate case where each multicast consists of only a single sender and a single receiver using a given QoS.

7.1.2.3 Single Point-to-Multipoint VC for Each Multicast Group
This case reduces even further the number of VCs by multiplexing all QoS reservations from the receivers on to a single VC. To ensure that all receivers receive at least the QoS requested in their respective resource reservations, this VC must be sized to provide resources to accommodate all the flows. The rules for accomplishing this could be similar to those used for merging of resource reservations in RSVP. For example, the PCR would be the maximum PCR among all receivers, the CTD would be the minimum, etc.

One disadvantage in this approach is that some receivers end up having more resources reserved than they actually requested. This could result in subsequent resource reservations being rejected unnecessarily because available resources were oversubscribed to other flows. Furthermore, best effort flows could also be unduly impacted. Thus, a practical implementation would suggest that at least two VCs be established: one for resource reservations and one for best effort flows.

There is also the possibility that a new or existing receiver may request a VC with QoS that cannot be supported by existing receivers (e.g., a PCR that exceeds the line rate of more receivers). In this case, multiple point-to-multipoint VCs must be maintained to support the flow.

7.1.2.4 Multicast Connection Management There are two multicast models supported by ATM. One is the multicast mesh, in which each sender establishes a point-to-multipoint VC to each receiver. The other model is the multicast server (MCS), in which each sender establishes a point-to-point VC to the MCS that establishes connectivity with each of the receivers in the multicast group for the distribution of multicast data. At the time of this writing, MARS does not support a means of communicating QoS information to the MCS. Therefore, the multicast mesh is the method for distributing multicast data under IIS.

7.1.3 Dynamic "Soft State" versus Static QoS

RSVP supports "soft state" reservations in which a receiver may alter the QoS of an established flow. ATM has no such mechanism for renegotiating QoS for an established VC. Thus, a change in a resource reservation means that a new VC may have to be set up and an old one torn down. In a point-to-multipoint VC multiple parties would be impacted.

The first issue to address is to determine in which order to perform activities. For instance, when setting up a replacement multipoint VC one can tear down the existing VC and begin transmitting on the new VC immediately. A drawback in this approach is that service is interrupted at a receiver for a period time from the termination of the current VC until the receiver is added to the new VC. Alternatively, if both the existing and new VC is maintained and if the flow is transmitted over both concurrently, some receivers will receive duplicate traffic. The best apparent solution would be to transmit only over the existing VC until the new VC is fully established. At that time, the sender ceases transmission over the old VC and sends subsequent traffic over the new one. The old VC can then be torn down.

There are a number of scalability concerns for the ATM network in this case. For instance, this could produce significant connection setup and teardown activity, which may exceed the capacity of the network if SVCs are used. Furthermore, there is significant potential for thrashing within the network. While the network is responding to one reservation change, another receiver in the same multicast group could issue an independent resource request, which obsoletes the VC in progress before it has been completed. Consider the following scenarios:

- A new VC is being established in response to an existing Receiver_A modified resource reservation when another receiver in the multicast group, Receiver_B, issues its own change in resource reservation that calls for greater (or less) resources than were requested by Receiver_A.

- A new VC is being established when a Receiver_C requests a resource reservation requiring VC resources that cannot be supported by some receivers in the current multicast group.
- As a new VC is being established in response to a modified resource reservation from an existing receiver, a new receiver joins the multicast group requesting a different QoS, or one not supported by some existing receivers.
- Existing receivers leave the multicast group.
- A sender may change its Tspec.

If the network operator has chosen to use an existing VC until a new one has been completed, the combination of these scenarios creates the possibility that an obsolete VC continues to be used simply because the volume of reservation changes doesn't allow the network enough time to complete a new one. This, in turn, could trigger retransmission of connection setup requests increasing even further the signalling traffic on the network.

The concern about VC completion could be addressed by flow controlling resource requests such that the network works on only one resource reservation at a time for a given multicast group, queuing all subsequent requests for service at a later time. The concern about signalling message retransmissions could be addressed by appropriate setting of time-out values in the signalling state machines within individual network elements.

7.1.4 Signalling Traffic

While it is important to ensure the appropriate QoS to RSVP data flows over ATM VCs, it is essential that provision be made for RSVP signalling messages. Since RSVP is based upon "soft state" information, there must be a means for edge devices to exchange refresh messages to maintain the standing reservations. Failing that, the reservation is released which may result in the associated ATM VC being torn down which would produce a negative impact on overall network QoS.

One possibility would be to use a best effort VC for the transmission of RSVP Path and Resv messages. This has the disadvantage that it doesn't address the need to transmit RSVP signalling messages even under network congestion conditions. A better solution would be to establish a fully connected mesh of point-to-point VCs for RSVP signalling traffic among all the edge devices in the network.

ATM protocols such as LANE and MPOA have time-out mechanisms that tear down SVCs that have been idle for measured time intervals. These mechanisms can potentially operate in conflict with RSVP. Such time-out mechanisms must be disabled when interworking with RSVP when SVCs are used.

7.1.5 Synchronizing ATM VC and RSVP States

An important aspect of mapping ATM VCs to RSVP flows is that the edge device must correlate state information between the two domains. Thus, if an Resv message triggers SVC establishment a confirmation of the result of the SVC establishment attempt should be returned to the receiver if a confirmation was requested in the Resv message. Furthermore, state information maintained in the edge device must be consistent between the ATM and RSVP domains. If an SVC establishment attempt is rejected within the ATM network, the edge device should clear RSVP state information for reservations that depend upon that SVC if an alternative SVC cannot be used. If an alternative SVC is used which cannot support the resource reservation, the RSVP state should also be cleared as this is to be treated as a best effort flow.

Likewise, if an existing VC should fail for any reason, the edge device must alter RSVP state information in accordance with the recovery action taken within the ATM network. Thus, a reroute to an alternative VC that is able to support the standing reservations should not result in a change of RSVP state, other than for the potential of temporary interruption of service between senders and receivers.

These RSVP state changes, whether induced by the ATM network or not, may result in changes with respect to how packets are processed in edge routers that employ WFQ and/or RED/WRED.

7.2 QoS Management

7.2.1 Guaranteed Service

Either CBR or rtVBR service categories may be used to support Guaranteed Service (GS) over ATM networks. CBR may, in fact, be used to support any IIS, however, the rtVBR maps more closely to the specifics of guaranteed service [17]. A mapping of equivalent parameters between rtVBR and the RSVP Resv message specifying guaranteed service is shown in Table 10-1.

Since guaranteed service provides best effort service for noncompliant packets, rtVBR with the cell tagging option should be used. If CBR service is used, however, PCR would be mapped to the RSVP R parameter in order to meet the delay bound required for GS. nrtVBR is not suitable for GS because of the lack of CTD and CDV specifications. ABR is unsuitable for this reason and because it does not provide burst tolerance. UBR is not suitable because it provides no QoS quarantees at all.

It should be noted that in these mappings between ATM and RSVP, actual ATM cell rates must be adjusted to account for AAL and ATM cell header overhead.

An ATM network may be characterized by the following RSVP delay parameters:

- Latency$_{ATM}$
- C_{ATM}
- D_{ATM}

Unless CBR is used, an estimation of C_{ATM} is very complicated. An acceptable approximation would be to model the ATM network as a pure delay element, in which case we could either ignore this term, or treat it as a first order effect that could be included in the estimate of D_{ATM}. Thus, an edge device may set the proper ATM CTD to support the desired end-to-end delay for GS according to the following relation:

$$CTD = D_{ATM} + S_{ATM} + Latency_{ATM}$$

where S_{ATM} represents the amount of slack allotted to the ATM network segment of the end-to-end flow route.

TABLE 10-1. rtVBR to GS mapping

rtVBR	GS
PCR	p
SCR	R
MBS	b

Since the actual queuing delay within an ATM network is variable, the variable portion of the CTD equation establishes an upper bound for CDV:

$$CDV \leq D_{ATM} + S_{ATM}$$

7.2.2 Controlled Load Service

CBR, rtVBR, or nrtVBR service categories may be used to support Controlled Load Service (CLS). nrtVBR maps more closely to the specifics of CLS, which has no delay bounds. A mapping of equivalent parameters between nrtVBR and the RSVP Resv message specifying CLS is shown in Table 10-2.

rtVBR may use this same mapping, but if CBR is used PCR must be greater than r to ensure that bursts are allowed to be emptied from the token bucket. The edge device is to provide sufficient buffering to accommodate packet bursts. ABR is not appropriate for CLS as it does not support packet bursts. If ABR is used, the MCR is to be set to a value no less than that of the r parameter. As with CBR, all buffering must be performed by the edge device. UBR is unsuitable for the reasons stated in the previous section.

7.2.3 Best Effort Service

Any ATM service category is adequate for best effort service, although UBR is the ATM service category that most closely maps to the IIS best effort service.

7.2.4 Traffic Tagging

It is always better for the edge device IWF to tag cells than for this to be done within the ATM network because the edge device has visibility into where the actual packet boundaries are, which is known to the ATM network only through the AUU information in the cell header.

Table 10-2. nrntVBR to CLS Mapping

nrtVBR	CLS
PCR	p
SCR	r
MBS	b

Even still, this information may be lost if network congestion causes the last cell in a packet to be lost. The edge device could determine which cells to tag based upon RED or WRED, which can be used to determine which packets are to be tagged during network congestion.

8. vBNS

The very high performance Backbone Network Service (vBNS) is an NSF sponsored Internet backbone network to support education and research institutions. The vBNS was activated on a test basis in late 1994 and went on-line as a production network in early 1995. The vBNS topology is shown in Figure 10-8.

The vBNS connects NSF supported Super Computer Centers (SCCs), and research institutions. At the time of this writing the vBNS was an ATM backbone that operated over OC-12 links and was beginning to deploy OC-48 links.

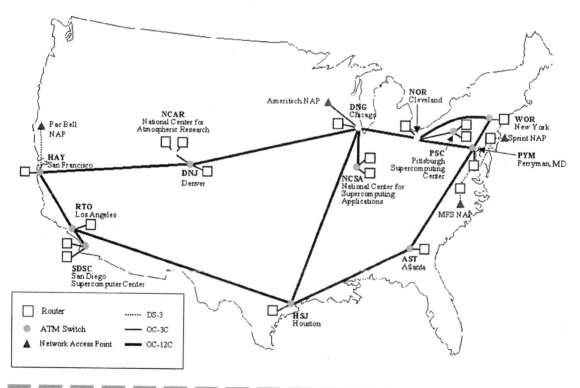

Figure 10-8. vBNS backbone network map.

Chapter 10: ATM and the Internet

The vBNS ATM configuration consists of a full mesh of UBR PVPs between ATM switches at vBNS nodes. Actual interconnection between nodes is via public carrier ATM network. vBNS ATM switches are colocated in the public carrier central office with the carrier's ATM backbone switches. vBNS nodes are connected to the Internet through the four NSF supported NAPs via DS-3 facilities. These PVPs become the bit pipes supporting a full mesh of PVCs among vBNS routers and provide transport for "best effort" traffic. In addition, a second set of VCCs is maintained for reserved bandwidth traffic. These VCCs are carried over a nrt-VBR PVP, with one such PVP established over each node-to-node OC-12 trunk shown in Figure 10-8. Consequently, reserved bandwidth flows between nodes without direct physical connection must exit the public carrier network where it is switched at the vBNS ATM switch to the next physical trunk en route to the destination.

An example of this strategy is shown in Figure 10-9. Connection VCC1 between vBNS ATM switches Switch3 and Switch2 is transported over the public carrier ATM network from over VPC1 from Switch3 to Switch1 which then switches VCC1 over VPC2 to the destination Switch2. Connection VCC2 is between "adjacent" nodes and is carried between Switch2 and Switch4 over VPC4.

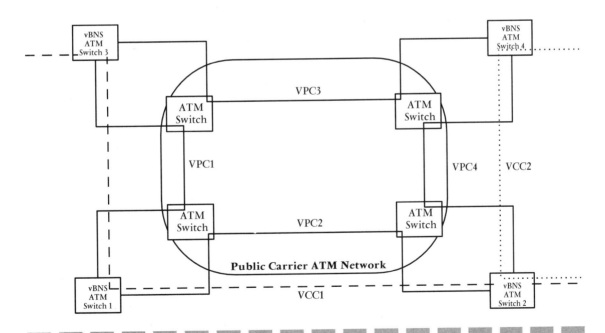

Figure 10-9. VP/VC strategy for reserved bandwidth connection.

The rationale behind the reserved bandwidth architecture in the vBNS is that it allows better use of the bandwidth available on the trunk. Thus, if needed, a connection could use the entire VPC bandwidth. By comparison, multiple VPCs in a fully connected mesh as is done with UBR VPCs would require the available link bandwidth to be divided up among the SCRs of individual VPCs on each physical OC-12 trunk. The obvious drawback is that a single connection could reserve all available SCR on a link but in that event, other connections would still have UBR connections available for use.

An advantage of the bandwidth management features of ATM in the vBNS is that network designers have considerable latitude in expanding the number of VPCs to support additional QoS classes, thereby segregating traffic by QoS requirement over the ATM network.

The current usage for the vBNS is for the carriage of data traffic. The reserved bandwidth channel is intended to be used to carry IP traffic using RSVP. This requires coordination between the RSVP and ATM procedures: RSVP admission and CAC must be coordinated such that they both accept and reject the same connections. RSVP reservations must also be mapped to UNI signalling parameters. Furthermore, the RSVP flow identifier must be mapped to the corresponding ATM VPI/VCI.

Since connections using the reserved bandwidth VPC are not intended to be permanent connections, an SVC capability is needed. This is accommodated through the use of ATM switches, which support endpoint-to-endpoint signalling, tunneled transparently through VPCs in the public carrier ATM network.

Attached to each vBNS node are the customers of the service—the education and research institutions that use the network. Figure 10-10 shows the architecture of an SCC site. Each SCC site consists of an ATM edge switch to which are attached two routers: one router supports FDDI attachment, while the other supports HIPPI attachment. The edge switch also supports direct ATM attachment. The edge switch then connects to the wide area network through direct attachment to a vBNS ATM switch.

9. ATM Deployment at Internet Network Access Points

Among the four NSF-sponsored NAPs, the Chicago and San Francisco NAPs have implemented ATM switches. Figure 10-11 shows the NAP at Chicago, Illinois. It is based upon an ATM switch which interconnects the routers and ATM switches belonging to the NSPs and ISPs which

Chapter 10: ATM and the Internet

Figure 10-10.
Supercomputing center vBNS access architecture.

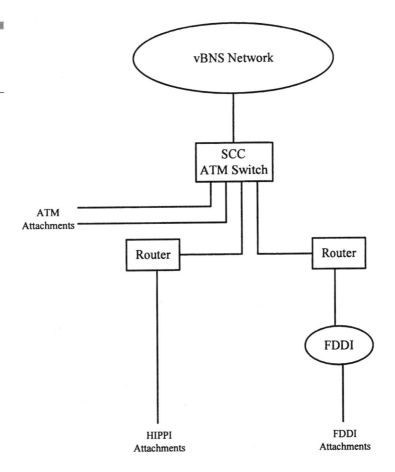

peer through the Chicago NAP. At the time of this writing interface rates supported at the Chicago ATM NAP ranged from DS-1 to OC-12. Interconnection through the NAP is by PVC providing UBR service.

The Chicago NAP is a layer 2 switched service and is not directly involved in routing of IP datagrams between ISPs. ISPs that interconnect at the NAP are assigned an IP address and a PVC assignment. It is the responsibility of the respective ISP/NSP routers to perform IP routing and IP to PVC translation. It is also the responsibility of peering ISPs to match AAL 5 encapsulation types across the PVC.

NSF-sponsored NAPs evolved in 1994. Consequently, decisions about technology deployed at those NAPs were made in that time frame. Opinions vary widely among the NAP operators as to which was the best suited technology for deployment with the New York and Washington DC NAPs choosing FDDI-based implementation.

Figure 10-11.
Chicago NAP architecture.

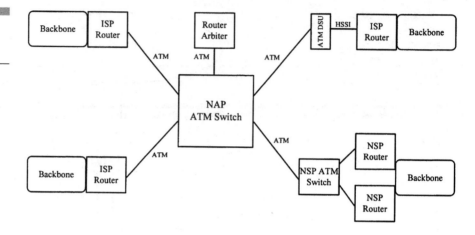

ATM was an early-stage technology during the mid-1990s time frame, and the NAP deployments faced the challenges inherent in the early adoption of an evolving technology. Performance problems in the early stages of deployment resulted in the ATM NAPs deploying parallel FDDI rings with hosts dual-homed on both the ATM switch and the local FDDI ring at the NAP.

Notwithstanding these challenges, the decision to deploy an ATM switch at the Chicago NAP was based upon the following considerations:

1. ATM provided scalability in port speeds that would allow the NAP to support increasing port speeds in response to customer demand on the same switch.

2. The ATM switch selected for deployment at the Chicago NAP was a nonblocking switching architecture that supported large numbers of subscribers. This would allow the NAP operator to support many subscribers and user ports on a single switch instead of having to deploy and maintain a matrixed arrangement of (comparatively) port-limited switches as port demand grew at the NAP.

The basic service provided at an NAP is transport interconnection between NSPs and ISPs which are peering through the NAP. The primary need is for large "pipes" to handle the increasing traffic volumes transferred across Internet backbones. Thus, any technology that can support large bandwidth pipes is adequate.

10. IP/ATM and IP/SONET

IP over SONET is a transport solution that provides a "pipe" between two connection endpoints. The question then becomes is there enough traffic to economically fill up the pipe since it is dedicated to point-to-point traffic.

IP over ATM (over a chosen transport medium such as SONET) is a switched solution, so traffic from a given pipe may be sent to many destinations.

An interesting aspect of the IP/SONET versus IP/ATM debate is that it gets back to the one-to-one versus one-to-many interconnection question. When frame relay services were introduced, a presumed impetus was that customers wanted to get away from dedicated point-to-point interconnection over private networks and toward a one-to-many solution where connectivity to many destinations could be achieved over a single interface into the public network. By comparison, the private line solution required a separate interface to each destination. In many ways the IP/SONET versus IP/ATM debate rehashes the same question under a different name.

10.1 Protocol Efficiencies

In this section we evaluate the efficiency of IP/SONET versus IP/ATM/SONET for the carriage of IP datagrams by comparing the overhead associated with each option. We assume an OC-3 SONET facility and a 1,500 octet IP datagram. We also assume that IP/SONET uses the Point to Point Protocol (PPP) with no padding [24].

10.1.1 SONET Overhead

Figure 10-12 shows the format of a SONET STS-3c Synchronous Payload Envelope (SPE) [22]. From Figure 10-12 we see that the actual payload capacity is 260 of the 270 octets for each of the 9 rows in the SPE. This yields a SONET payload efficiency of 96.3%.

10.1.2 IP/PPP/SONET Overhead

We assume that IP datagrams are transmitted over SONET using the Multiple Access Protocol over SONET/SDH (MAPOS) [19][20]. In this

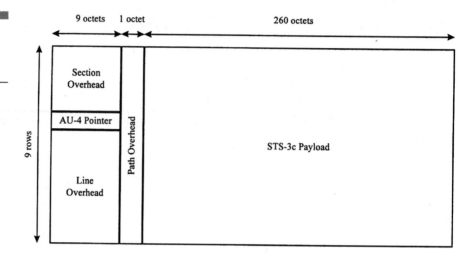

Figure 10-12.
SONET STS-3c synchronous payload envelope.

case, IP/SONET overhead consists of 7 octets added to the 1,500 octets of information payload in the IP datagram. PPP overhead is 2 octets. This yields an IP/PPP/SONET payload efficiency of 99.4%.

Thus the total payload efficiency of IP/SONET is:

$$96.3\% \times 99.4\% = 95.7\%$$

which, on a 155.52 Mbits/s SONET OC-3 link, translates into a transfer rate of 148.8 Mbits/s.

10.1.3   IP/ATM/SONET Overhead

We assume AAL 5 with LLC/SNAP encapsulation [21]. LLC/SNAP encapsulation overhead is 8 octets. AAL 5 adds an additional 8 octets of overhead plus 20 octets of padding (necessary to make the AAL 5 PDU size an integer number of ATM cells). This yields an LLC/SNAP/AAL 5 payload efficiency of 97.7%.

After segmentation the AAL 5 PDU is divided into ATM cells which contain 5 octets of overhead and 48 octets of payload. Thus the ATM layer payload efficiency is 90.6% which yields an IP/ATM payload efficiency of

$$90.6\% \times 97.7\% = 88.5\%$$

Thus the total IP/ATM/SONET efficiency is 85.2%—a SONET OC-3 transfer rate of 132.5 Mbits/s.

10.1.4 Comparative Efficiencies IP/PPP/SONET versus IP/ATM

From the above analysis we can see that IP/SONET provides approximately 10% better efficiency than IP/ATM. In 10-3 Table we present a comparison of protocol efficiencies, and effective transfer rate (for a SONET OC-3 link) for IP datagrams of 1,500 octets, 576 octets, and 44 octets. From Table 10-3 we see that IP/ATM protocol efficiency can fall dramatically relative to IP/SONET. This is primarily due to the amount of padding required by AAL 5 relative to the datagram size. Thus, protocol efficiency is less sensitive to padding for large datagram size, but protocol efficiency can be extremely sensitive to pad insertion for small datagram size. At 1,500 octets, padding reduces ATM protocol efficiency by only about 1%, but at 44 octets padding reduces protocol efficiency by nearly 25%.

TABLE 10-3. Protocol efficiency/effective data transfer rate comparison

IP Datagram Size (octets)	IP/PPP/SONET		IP/ATM/SONET	
	Efficiency	Transfer Rate	Efficiency	Transfer Rate
1,500	95.7%	148.8 Mbits/s	85.2%	132.5 Mbits/s
576	94.8%	147.4 Mbits/s	80.5%	125.2 Mbits/s
44	89.9%	139.8 Mbits/s	40%	62.2 Mbits/s

10.1.5 Empirical Results

The 1997 study of Internet traffic over an NSP backbone cited earlier in this chapter used the actual traffic patterns observed to estimate the relative efficiencies of IP/SONET and IP/ATM for the observed traffic profile [6]. Based upon 16,794,689,797 observed packets the following bit-efficiency comparison was made between IP/ATM and IP/SONET:

TABLE 10-4. Comparison between IP/ATM and IP/SONET. Empirical results of study

	ATM	SONET
IP payload octets	5,777615,952,291	5,777615,952,291
Total octets	7,233,116,621,345	6,121,916,426,350
Bit efficiency	80%	94%

10.2 IP/SONET versus IP/ATM: Which Is Better?

There is no way to prove that one is "better" than the other, so no attempt will be made here to do any such thing. At face value, the question of whether IP/SONET is better than IP/ATM is somewhat of an "apples and oranges" comparison—like asking which is better, transport or switching? ATM or SONET?

From the end-user perspective, IP/SONET is a transport solution. In this case, the end user gets a bit pipe with all the traffic in that interface being shipped point to point to a destination. The question from the end-user perspective then becomes one of whether the end user has enough traffic going to that destination to economically dedicate an interface to the delivery of that traffic. ATM, however, is a switching solution. Therefore, traffic being carried over a single interface may be switched to any number of destinations.

When looking at this question from the core network operator perspective, other pieces need to be added to the discussion to make this question sensible. While IP/ATM is a switched solution, a core network deployment of IP/SONET would implement "switching" functionality in routers which terminate the point-to-point SONET links and forward IP datagrams to the next SONET link along an end-to-end route. Historically, switches have had a performance edge in forwarding of packets in comparison to routers. Layer 3, however, switches perform in hardware many of the functions historically performed in software by traditional routers. This has produced a generation of IP datagram forwarding engines that perform switching at rates approaching media speed [23]. An advantage enjoyed by ATM switches was that they supported higher rate interfaces and greater switching capacities than routers. However, the emergence of high capacity routers has narrowed this edge, making sheer bandwidth a less significant metric of comparison.

Thus, a network of routers connected by IP/SONET links begins to appear topologically equivalent to a network of ATM switches. Ultimately, what constitutes the "better" choice between IP/SONET versus IP/ATM depends upon one's perspective on the importance of guaranteed Quality of Service. ATM supports mechanisms for end-to-end quality of service guarantees. At this writing, the primary QoS mechanisms available on routers, that support it, provide QoS on a link-by-link basis only. End-to-end QoS in routers would require enhancements like RSVP which do not actually guarantee end-to-end QoS since confirmation of a resource reservation is only made between the destination and the nearest heterogeneous branch point. Thus, a network operator will have to determine what constitutes a "good enough" level of service when making a technology decision. The importance of the "QoS factor" is also dependent upon choices about how network bandwidth is to be managed (as we saw in the vBNS network), as well as the network operator's beliefs about the likelihood of an emergence of demand for high bandwidth, isochronous services (such as multimedia).

The obvious observation about IP/SONET in comparison to IP/ATM is that IP/SONET is a more efficient means of transporting IP datagrams than IP/ATM/SONET. In addition, as we saw in the last chapter, transport of TCP (or UDP) over ATM provides no inherent advantages over packet switching networks, which discard whole packets and therefore do not waste resources on delivery of particle packets. Packet switching networks are optimized for the delivery of packet based traffic. The degree that ATM is able to provide advantages in comparison to IP/SONET is to the extent that supported service classes and end-to-end QoS guarantees of ATM are critical to the management of bandwidth, or to the types (and levels) of services that are to be offered over the network.

References

1. Nelson, David, ATM in the Service Provider Market, Telecommunications, March 1998.
2. Comer, Douglas, *Internetworking with TCP/IP 2d Ed.*, Prentice-Hall, 1991.
3. Malkin, G., The Tao of IETF—A Guide for New Attendees of the Internet Engineering Task Force, Internet Engineering Task Force, RFC 1718, November 1994.
4. Claffy, K. G., Polyzos, and H-W. Braun, Traffic Characterizations of the T1 NSFNET Backbone, Proceedings of INFOCOM '93, San Francisco CA, March 1993.
5. Frazer, K. D., NSFNET: A Partnership for High-Speed Networking, Final Report 1987—1995, Merit Network, Inc., 1995.
6. Thompson, Kevin, Gregory Miller, and Rick Wilder, Wide-Area Internet Traffic Patterns and Characteristics, IEEE Network, November/December 1997.
7. Cerf, V., and R. Kahn, A Protocol for Packet Network Intercommunication, IEEE Transactions on Communications, Vol 22, No. 5, May 1974.

8. R. Braden, D. Clark, and S. Shenker, Integrated Services in the Internet Architecture: an Overview, Internet Engineering Task Force, RFC 1633, June 1994.

9. S. Floyd, and V. Jacobson, Random Early Detection Gateways for Congestion Avoidance, IEEE/ACM Transactions on Networking, August 1993.

10. Braden, B. et. al., Recommendations on Queue Management and Congestion Avoidance in the Internet, Internet Engineering Task Force, RFC 2309, April 1998.

11. Deering, S, and R. Hinden, Internet Protocol, Version 6 (IPv6) Specification, Internet Engineering Task Force, RFC 1883, December 1995.

12. Braden, Bob, Lixia Zhang, Steve Berson, Shai Herzog, and Sugih Jamin, Resource Reservation Protocol (RSVP)-Version 1 Functional Specification, Internet Engineering Task Force, RFC 2205, September 1997.

13. Shenker, Scott, Craig Partridge, and Roch Guerin, Specification of Guaranteed Quality of Service, Internet Engineering Task Force, RFC 2212, September 1997.

14. Wroclawski, John, Specification of the Controlled-Load Network Element Service, Internet Engineering Task Force, RFC 2211, September 1997.

15. Wroclawski, John, The Use of RSVP with IETF Integrated Services, Internet Engineering Task Force, RFC 2210, September 1997.

16. Crawley, Eric, et. al., A Framework for Integrated Services and RSVP over ATM, Internet Engineering Task Force, Internet Draft, May 1998.

17. Garrett, Mark, and Marty Borden, Incorporation of Controlled-Load Service and Guaranteed Service with ATM, Internet Engineering Task Force, Internet Draft, March 1998.

18. The Emerging Digital Economy, United States Department of Commerce Report, April 1998.

19. Murakami, Ken, and Mitsuru Maruyama, MAPOS—Multiple Access Protocol over SONET/SDH Version 1, Internet Engineering Task Force, RFC 2171, June 1997.

20. Murakami, Ken, and Mitsuru Maruyama, IPv4 over MAPOS Version 1, Internet Engineering Task Force, RFC 2176, June 1997.

21. Heinanen, Juha, Multiprotocol Encapsulation over ATM Adaptation Layer 5, Internet Engineering Task Force, RFC 1483, July 1993.

22. Synchronous Optical Network (SONET)—Basic Description including Multiplex Structure, Rates, and Formats, American National Standards Institute, ANSI T1.105, 1995.
23. Tolly, Kevin, John Curtis, and Andy Hacker, Straight Talk on Layer 3 Switching, Business Communications Review, September 1997.
24. Simpson, William, The Point-to-Point Protocol, Internet Engineering Task Force, RFC 1661, July 1994.

CHAPTER 11

Voice Over ATM

1. Introduction

The carriage of voice traffic over ATM networks is an important part of the vision of an integrated network capable of replacing special purpose networks. In this chapter we will bring together a number of topics covered in previous chapters in a discussion of voice, and other real time services such as multimedia, over ATM networks.

We will begin with a discussion of voice traffic carried directly over ATM networks. In this treatment, voice traffic is carried directly from a Broadband TE (B-TE) across an ATM network where it is interworked to the narrowband ISDN network.

From there we will discuss the emerging topic of voice over the Internet as defined in the ITU H.323 recommendation. In this part of the chapter we will discuss how voice and multimedia are to be carried over the Internet. Finally, we will discuss the integration of H.323 and ATM networks.

2. Voice over ATM to the Desktop

A reference configuration for connection of a B-TE on an ATM network to an endpoint on a narrowband ISDN network (TE) is shown in Figure 11-1 [1]. Traffic carried over the connection is 64 kbps PCM-encoded voice as described in ITU recommendation G.711 [2]. The protocol reference model for interworking of User Information and Signalling traffic is also shown in Figure 11-1. Each voice call from the B-TE uses one VCC in the ATM network.

B-TEs attached to a private ATM network have the option of using either AAL 1 or AAL 5. Since signalling requires use of AAL 5, B-TE will typically support AAL 5, which is also the most efficient AAL for data applications. However, a B-TE attached to a public ATM network may be required to use AAL 1, which is a more efficient protocol for carrying small voice packets than is AAL 5. This creates the possibility that a call originated at the B-TE using AAL 5 on a private ATM network may be transcoded into AAL 1 for transport over a transit public ATM network before delivery across a second IWF to a N-ISDN network for subsequent delivery to the destination, as shown in Figure 11-2.

PDU size for voice traffic is driven by delay considerations. Thus, voice packets would not be expected to be large. As we have seen earlier, AAL 5 efficiency is sensitive to packet size. So for small packet size, it is

Chapter 11: Voice Over ATM

Figure 11-1.
Voice over ATM interworking configuration.

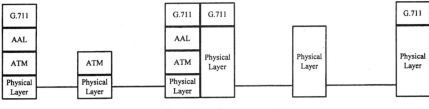

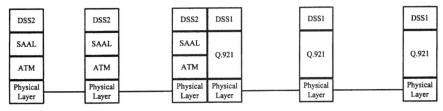

essential to pick a packet size that is optimal for AAL 5 to eliminate padding octets. The objectives of minimized delay and elimination of padding can be achieved by picking a packet size such that payload and AAL trailer are able to fit into a single ATM cell without padding. For voiceband information the AAL 5 CPCS-PDU payload may be up to 40 octets. For voice rate traffic, the cell rate is 200 cells/sec assuming fully filled cells. For 64 kb/s traffic, samples arrive at a rate of 1 octet every 125 μs, so accumulation of the 40 voice samples needed to fully fill the cell introduce a cell construction delay of 5 ms.

The B-ISDN to N-ISDN IWF performs signalling interworking by translating broadband access signalling (DSS2) messages to equivalent narrowband access signalling (DSS1) messages as shown in Figure 11-1 [3][4]. The signalling protocol reference model shown in the Figure 11-1 applies

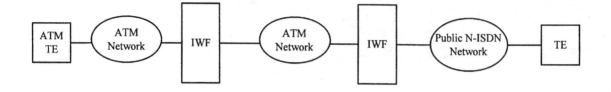

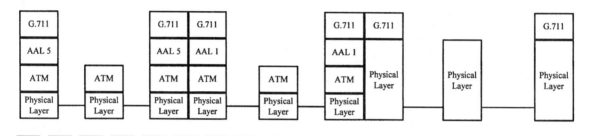

Figure 11-2. AAL 1 to AAL 5 transcoding in voice path.

to access signalling between a private ATM network and a public narrowband ISDN network across a public UNI. Interconnection between a public ATM network and a public N-ISDN network across a B-ICI requires procedures for interworking between B-ISUP signalling in the ATM network, and ISUP signalling in the N-ISDN network [5][6]. At the time of this writing, these procedures have not been specified in the ATM Forum B-ICI specification but have been by the ITU [7] [16].

The IWF(s) may also be required to perform address translation from a non-E.164 addressing scheme used in the private ATM network, to E.164 addressing used in the public network.

3. Recommendation H.323—Voice over the Internet

At the time of this writing, an important emerging standard for voice over the Internet was based upon ITU Recommendation H.323, which is

Chapter 11: Voice Over ATM

a protocol suite that defines procedures for multimedia communications services over packet based networks which do not provide guaranteed quality of service [8]. The implications for voice carried directly over ATM as described in the last section are unclear, but the ability for ATM networks to interwork with H.323 networks is an important area of investigation at the time of this writing.

As shown in Figure 11-3, an H.323 network includes the following elements:

- Endpoints, which include H.323 entities such as terminals and gateways.
- Terminals—devices that provide for real-time, two-way multimedia communications with other H.323 endpoints.

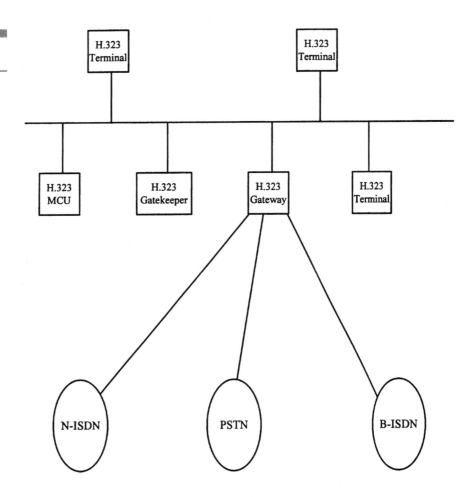

Figure 11-3.
H.323 topology.

- Gateways — devices that provide for real-time, two-way communications between H.323 terminals on the packet network and terminals on a N-ISDN, B-ISDN, or PSTN.
- Gatekeepers — devices that control access to the packet network by endpoints and may provide other services such as address translation between packet network addresses and PSTN addresses and bandwidth management.
- Multipoint Control Units (MCU) — devices that enable three or more terminals and gateways to participate in a conference. The MCU is the multimedia equivalent of the conference bridge used in telephony. The MCU consists of a multipoint controller (MC) which provides control functions in the conference, and optionally a multipoint processor (MP) which processes audio and/or video performing functions such as mixing, switching, and other media processing functions.

A collection of terminals, gateways, and MCUs may be associated with a gatekeeper that controls the ability of the endpoints to communicate. This collection of endpoints and a gatekeeper is known as a *zone*.

In the following description of H.323 we will assume IP at the network layer and TCP/UDP at the transport layer.

3.1 Description of H.323

Prior to any H.323 communication, endpoints register with a gatekeeper. Communication between the gatekeeper and endpoints takes place over a Registration, Admissions, and Status (RAS) channel as shown in Figure 11-4. Endpoints register by sending a Registration Request (RRQ) message to the gatekeeper over the RAS channel. The RAS channel is identified by an IP address and a UDP port. The RAS channel, therefore, uses *unreliable* transport with no acknowledgment of messages between sender and receiver being provided by the UDP transport layer protocol. The (IP address, TCP/UDP port) tuple will henceforth be referred to as a *transport address* for the entity. The gatekeeper RAS channel transport address is either known by the endpoint based upon configuration, or may be discovered using a Gatekeeper Discovery process. The gatekeeper discovers the endpoint's RAS channel transport address through the registration process. If the registration request is accepted by the gatekeeper a Registration Confirmation (RCF) message is sent back to the endpoint indicating admittance of the endpoint to the gatekeeper's zone. At this point, the endpoint is able to communicate using H.323 procedures.

An H.323 communication includes the following phases:

1. Call Setup
2. Initial Communication and Capability Exchange
3. Establishment of Audiovisual Communication
4. Call Termination

3.1.1 Call Setup

Figure 11-5 shows the message exchange that takes place during the first phase of communication originated by Endpoint 1. Endpoint 1 initiates the process by sending an Admission Request (ARQ) message to the gatekeeper over the RAS channel to request resources to communicate with a destination. If the gatekeeper accepts the ARQ it returns an Admission Confirmation (ACF) to Endpoint 1 which contains the transport address to be used for the establishment of an H.225 call-signalling channel [9]. In Figure 11-5, the returned address is the address of Endpoint 2. If the transport address of Endpoint 2 is the transport address of the destination, then the call is to be established between the two endpoints. If not, Endpoint 2 may be a gateway that provides interworking between Endpoint 1 and a called destination on another network.

Messages transported over the H.225 call-signalling channel are a derivative of those used in Q931/DSS 1 procedures. The H.225 call-signalling channel is identified by an IP network layer address and a TCP port. The H.225 call-signalling channel, uses *reliable* transport with messages sent being acknowledged back to the sender by the TDP transport layer protocol. Endpoint 1 sends a SETUP message to Endpoint 2 to establish a

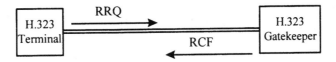

Figure 11-4.
H.323 endpoint registration.

Figure 11-5.
H.323 call setup.

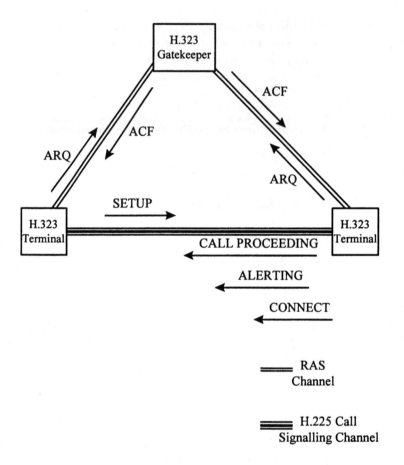

connection between the endpoints. In response, Endpoint 2 sends a CALL-PROCEEDING message back to Endpoint 1 over the call-signalling channel to inform Endpoint 1 that the SETUP message was received and is being processed. The transport address at Endpoint 1 for the H.225 call-signalling channel is sent with the SETUP message. Next, Endpoint 2 executes an ARQ/ACF exchange with the gatekeeper over the RAS channel.

If Endpoint 2 is the called destination, an ALERTING message is sent to Endpoint 1 over the H.225 call-signalling channel with a subsequent CONNECT message when the call is answered.

If Endpoint 2 is a gateway, it provides interworking of the H.323 SETUP message to an appropriate message, which is sent over a call-signalling channel in the external (ATM, N-ISDN, PSTN, etc.) network. As call-related messages are received from the external network, they are interworked at the gateway into corresponding H.323 messages (ALERTING, CONNECT, etc.) which are sent to Endpoint 1.

3.1.2 Initial Communication and Capability Exchange

Once a communication has been established between endpoints, an H.245 Control Channel is established [10]. Optionally, this channel may be established prior to receipt of an H.323 CONNECT message (such as after an H.323 SETUP/CALL-PROCEEDING message exchange). The purpose for this channel is to allow the endpoints to negotiate the multimedia capabilities that each endpoint will use in the communication. The H.245 control channel is identified by an IP network layer address and a TCP port. The transport addresses used for the H.245 control channel are discovered during the H.323 SETUP/CONNECT message exchange.

During this phase, the endpoints exchange H.245 terminalCapabilitySet messages as shown in Figure 11-6. These messages describe terminal capabilities, such as the ability to receive G.711 encoded audio or H.261 encoded video.

A gateway must interwork the contents of the terminalCapabilitySet messages to an end-to-end negotiation with a destination in the external network.

3.1.3 Establishment of Audiovisual Communication

After negotiating terminal capabilities, the endpoints establish a media channel that is to carry the audio, video, or data traffic between the originating and terminating endpoints in the communication.

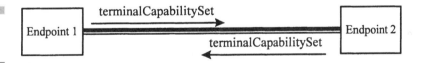

Figure 11-6. Terminal capabilities exchange.

This exchange, shown in Figure 11-7, is affected by an OpenLogicalChannel/ OpenLogicalChannelAck exchange over the H.245 control channel. These messages contain the transport addresses that are to be used for the media channel as well as providing a set of parameters that describes the capabilities required in the channel. For data traffic TCP/IP is used, while for audio and video streams UDP/IP is used.

The media channel used for audio and video streams within H.323 is based upon the Real Time Protocol (RTP) [11]. RTP is a transport protocol for real-time applications. It supports these applications by providing time stamp and packet sequence number information in the RTP header. The actual audio or video stream (which is encoded using a method defined during the capabilities exchange) is encapsulated

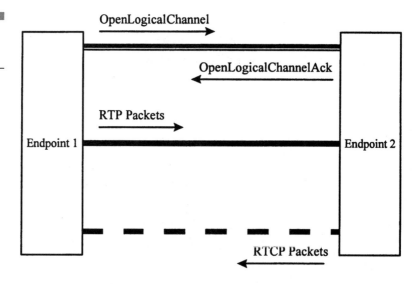

Figure 11-7.
Establishment of the RTP media channel.

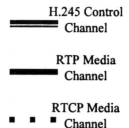

within the RTP packet. RTP also supports multipoint communications by providing header information that identifies the stream source and any multimedia mixing devices that have altered the stream en route to the destination.

The basic RTP header is shown in Figure 11-8. It consists of the following fields:

- Version (V): a 2-bit field that identifies the RTP version.
- Padding (P): a 1-bit field that indicates whether there are padding octets at the end of the packet.
- Extension (X): a 1-bit field that indicates whether the packet has an extended header. RTP supports the definition of application-dependent header extensions.
- Contributing Source Count (CC): a 4-bit field that indicates the number of Contributing Sources (CSRCs) included in the packet. When traffic from multiple sources is mixed into a single stream at an MP, each of the previously independent sources becomes a CSRC to the mixed stream, which is the output of the MP.
- Marker (M): a 1-bit field whose use is application dependent. In general, the M bit is used to indicate a significant boundary such as the last packet of a multipacket frame.
- Payload Type (PT): a 7-bit field that identifies the format of the RTP payload.
- Sequence Number: a 16-bit field that increments by one for each RTP packet sent.
- Time stamp: a 32-bit field that contains a time stamp for the RTP packet.
- Synchronizing Source (SSRC): a 32-bit field that identifies an RTP media stream source.
- CSRC List: a list consisting of 0 to 15 32-bit entries that identifies the RTP sources that contributed to the media stream.

RTP does not provide QoS guarantees; however, a companion protocol, the RTP Control Protocol (RTCP) is based upon the periodic transmission of control packets to provide quantitative feedback on the quality of service being rendered over the RTP media channel.

RTP and RTCP may be carried over separate channels as determined during the OpenLogicalChannel message exchange over the H.245 control channel.

3.1.4 Call Termination

Figure 11-9 shows H.323 call termination. When terminating an H.323 call, an endpoint is to send an EndSessionCommand over the H.245 control channel to terminate the media channel. The H.245 control channel is itself closed upon receipt of an EndSessionCommand from the peer endpoint.

Next, the endpoint sends a RELEASE COMPLETE message over the H.225 call-signalling channel to close the channel.

Each H.323 endpoint clears the call by sending a Disengage Request (DRQ) message to the gatekeeper. The gatekeeper responds with a Disengage Confirmation (DCF) message. The RAS channel remains open as long as the endpoint has membership within a given zone.

4. H.323 Over ATM

By using VCCs over an ATM network instead of a packet network, H.323 endpoints are able to establish QoS-based media streams. The most straightforward method of implementing H.323 over ATM would be to use an IP over ATM method, such as the ones described

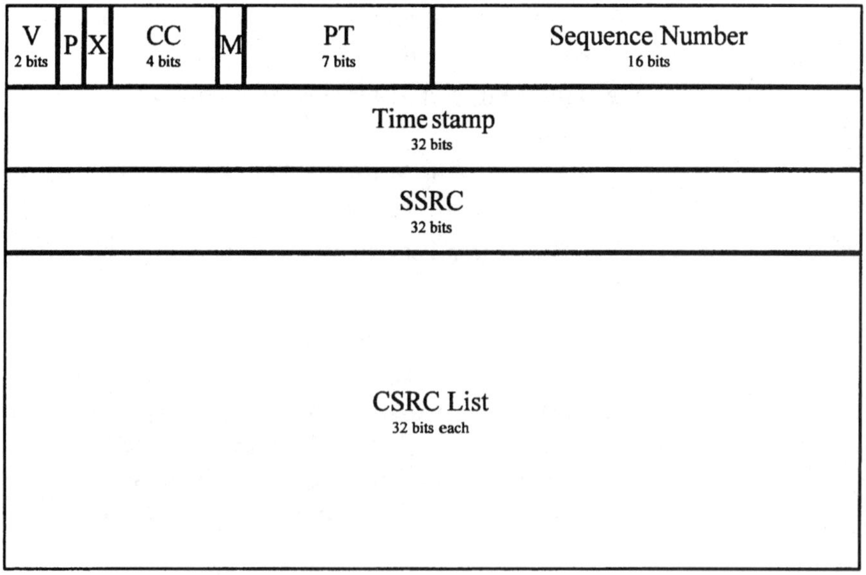

Figure 11-8.
RTP header.

V 2 bits	P	X	CC 4 bits	M	PT 7 bits	Sequence Number 16 bits
Time stamp 32 bits							
SSRC 32 bits							
CSRC List 32 bits each							

Chapter 11: Voice Over ATM

Figure 11-9.
H.323 call termination.

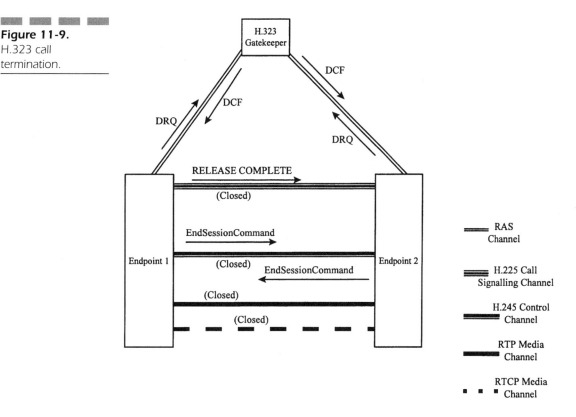

in Chapter 8. This approach is less efficient than carrying the RTP media streams directly over ATM, however. Since packets may be of variable size, AAL 5 is used instead of AAL 1 for RTP over ATM. Regardless of the media channel used, H.323 requires that all control channels be carried over IP (or a packet network protocol, in general) to ensure interoperability between ATM-enabled H.323 terminals, and those which do not. This still allows those channels to be carried through an IP over ATM method, but only the RTP media channel is allowed to be modified for an underlying ATM network technology.

Figure 11-10 shows how the setup for an H.323 media channel over ATM differs from that of an H.323 over IP setup [12]:

1. The OpenLogicalChannel exchange still takes place between Endpoint 1 and Endpoint 2 over the H.245 control channel. In this case, the message from the initiating endpoint (which we will assume to be Endpoint 1) will contain parameters for an ATM media

Figure 11-10.
Establishment of RTP media stream over ATM.

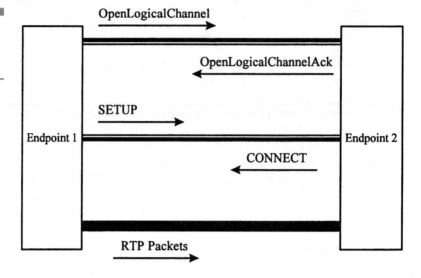

connection. The parameters in this message contain information that will be used by the receiver to set up an ATM connection.

2. When Endpoint 1 receives the OpenLogicalChannelAck, it sets up an ATM SVC. The ATM SETUP message describes the RTP encapsulation header in the *Broadband Low Layer Information* element. In addition, however, there must be a means of correlating the RTP media stream address to the address of the ATM VCC (which can be considered as an RTP to ATM address resolution). This is accomplished by including a connection identifier, such as an RTP port number for the media stream, in the *Broadband High Layer Information* element of the ATM SETUP message. If Endpoint 1 is an H.323 gateway, the OpenLogicalChannelAck is held until the ATM connection is established. For media streams over ATM, either CBR or nrt-VBR service classes may be used.

3. Receipt of an ATM CONNECT message, informs Endpoint 1 that the RTP media channel from Endpoint 1 to Endpoint 2 is now open over the ATM network. If Endpoint 2 is a gateway, the OpenLogicalChannelAck is forwarded to the call-originating endpoint thereby informing the call originator that the RTP media channel has been opened.

The H.323 "Fast Connect" procedure simplifies ATM SVC establishment. Fast Connect allows the call-initiating endpoint (Endpoint 1) to include a fastStart element in the H.323 SETUP message, sent over the H.245 call-signalling channel, which contains a series of OpenLogicalChannel structures that describe proposed media to be used for the exchange of media streams between the endpoints involved in the communication. The receiving endpoint (Endpoint 2) accepts the Fast Connect procedure by including a fastStart element in an H.323 message sent in response to the SETUP message. This return fastStart element contains the media channels that have been accepted for media communication. In this case, Endpoint 2 initiates establishment of an ATM VCC. If Endpoint 2 is a gateway, it suspends return of an H.323 ALERTING, or CONNECT message toward Endpoint 1, pending establishment of an ATM SVC. Upon receipt of an ATM CONNECT message, the H.323 ALERTING/CONNECT message is sent to Endpoint 1. Establishment of an ATM connection using the Fast Connect procedure is shown in Figure 11-11. As can be seen in Figure 11-11, the Fast Connect procedure allows an ATM media stream connection to be set up in as little as one bidirectional message exchange over the H.225 call-signalling channel.

4.1 Protocol Efficiency

H.225 defines the default packetization interval to be 20 ms for packetized audio. The actual packetization interval used over a media channel is negotiable. Assuming a 20ms packetization interval and a G.711 codec, this produces a packet size of 160 octets. RTP encapsulation adds at least 12 octets of header. UDP encapsulation adds another 8 octets of header, IP adds at least 20 octets of header, and PPP adds 2 octets of overhead [13]. The total overhead is at least 42 octets which produces an RTP/UDP/IP/PPP protocol efficiency of at most 79.2%. RTP/UDP/IP/ATM produces a protocol efficiency of 60.4% when using LLC/SNAP encapsulation and AAL 5.

By comparison, the protocol efficiency for RTP/ATM is 81.2%. This is slightly more efficient than carrying RTP/UPD/IP/PPP, and is also more

Figure 11-11.
Establishment of RTP media stream over ATM using fast connect procedure.

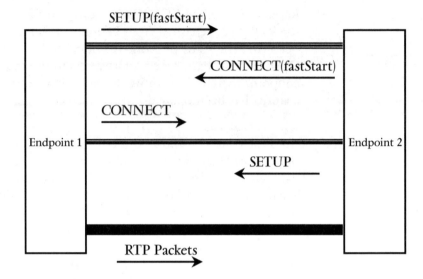

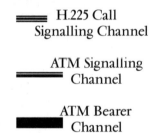

efficient than the 75.5% protocol efficiency realized when carrying voice directly over ATM using AAL 5 with 40 octets of voice payload. All of these options are less efficient than carrying voice over AAL 1 which yields 88.7% protocol efficiency for a fully utilized 47 octet voice payload.

These results are summarized below:

	RTP/UDP/IP/ PPP	RTP/UDP/IP/ ATM	RTP/ATM	Voice/ATM (AAL 5)	Voice/ATM (AAL 1)
Voice Payload Size	160 octets	160 octets	160 octets	40 octets	47 octets
Protocol Efficiency	79.2%	60.4%	81.2%	75.5%	88.7%
Cell Rate	—	250 cells/s	200 cells/s	200 cells/s	171 cells/s

4.2 Interworking H.323 and Non-H.323 Terminals

When interworking a B-TE supporting voice over ATM to an H.323 terminal, an H.323 gateway is required. The gateway must remove the RTP/UDP/IP headers in the H.323 to B-TE direction, and generate a RTP/UDP/IP header in the opposite direction. If the B-TE uses AAL 1, an IWF must also transcode between AAL 1 and AAL 5. Furthermore, if the H.323 terminal uses voice compression techniques such as those specified in G.728 (16 kb/s voice), or G.729 (8 kb/s voice), additional delays are incurred due to transcoding between G.711 and G.728/G.729 at the IWF [14][15].

References

1. Voice and Telephony Over ATM to the Desktop Specification, ATM Forum, Document AF-VTOA-0083.000, May 1997.

2. Pulse Code Modulation (PCM) of Voice Frequencies, International Telecommunication Union, Recommendation G.711, 1993.

3. Digital Subscriber Signalling System No. 1 (DSS 1)—ISDN User-Network Interface Layer 3 Specification for Basic Call Control, International Telecommunication Union, Recommendation Q.931, 1993.

4. Broadband Integrated Services Digital Network (B-ISDN)—Digital Subscriber Signalling System No. 2 (DSS 2)—User-Network Interface (UNI) Layer 3 Specification for Basic Call/Connection Control, International Telecommunication Union, Recommendation Q.2931, 1995.

5. Signalling System No. 7 ISDN User Part (ISUP)—Basic Call Procedures, International Telecommunication Union, Recommendation Q.764, 1993.

6. Broadband Integrated Services Digital Network (B-ISDN)—Signalling System No. 7 B-ISDN User Part (B-ISUP)—Basic Call Procedures, International Telecommunication Union, Recommendation Q.2764, 1995.

7. BISDN Inter Carrier Interface (B-ICI) Specification Version 2.0 (Integrated), ATM Forum, Document af-bici-0013.003, December 1995.

8. Packet-Based Multimedia Communications Systems, International Telecommunication Union, Recommendation H.323, 1998.

9. Call Signalling Protocols and Media Stream Packetization for Packet Based Multimedia Communications Systems, International Telecommunication Union, Recommendation H.225.0, 1998.

10. Control Protocol for Multimedia Communication, International Telecommunication Union, Recommendation H.245, 1997.

11. Schulzrinne, Henning, Stephen Casner, Ron Frederick, and Van Jacobson, RTP: A Transport Protocol for Real-Time Applications, Internet Engineering Task Force, RFC 1889, January 1996.

12. H.323 Media Transport Over ATM, ATM Forum, Document BTD-SAA-RMOA-01.04, October 1998.

13. Internet Protocol DARPA Internet Program Protocol Specification, Defense Advanced Research Projects Agency, IETF RFC 791, September 1981.

14. Coding of Speech at 16 kbit/s Using Low-Delay Code Excited Linear Predication, International Telecommunication Union, Recommendation G.728, 1992.

15. Coding of Speech at 8 kbit/s Using Conjugate-Structure Algebraic-Code-Excited Linear-Prediction (CS-ACELP), International Telecommunication Union, Recommendation G.729, 1996.

16. Broadband Integrated Services Digital Network (B-ISDN)—Interworking Between Signalling System No. 7—Broadband ISDN User Part (B-ISUP) and Narrowband ISDN User Part (N-ISUP), International Telecommunication Union, Recommendation Q.2660, February 1995.

PART 4

CHAPTER 12

Future Directions for ATM

1. Introduction

ATM has been a technology that has, to a degree, been in search of an application. It first burst on the scene with great fanfare, maybe too much fanfare, in the early 1990s as a solution to bringing massive volumes of bandwidth to the desktop. At present, it is viewed as a technology that has been deployed largely within public carrier networks. Ad interim ATM has suffered from a loss of credibility as a result of too much being promised too soon about the capabilities and potential of the technology. It should be added, however, that an overdose of optimism appears to be the greeting for practically every new "hot" communications technology in the current environment.

But while ATM burst on the scene as a solution to higher bandwidth, its primary advantages are in the management of bandwidth. This capability allows different types of traffic which are currently carried on separate overlay networks to be carried over a common network. We have seen an example where ATM is used in an Internet backbone to provide for classification of Internet traffic—some traffic receiving "best effort" delivery, with other traffic receiving reserved bandwidth for higher priority delivery.

1.1 Managed Bandwidth versus Infinite Bandwidth

One of the drivers for ATM technology was the belief that as network bandwidth increased, and as different types of information (voice, video, and data) were merged into a common network from previously existing overlay networks, there would be a need to manage the bandwidth to deliver service quality levels to meet the requirements of diverse applications.

An alternative school of thought has been that by providing enough bandwidth in networks, it would be possible to minimize queuing delays, and to increase the information-handling capacity of the networks such that bottlenecks effectively disappear. Thus, information transfers would be able to occur at "media speed" with minimum end-to-end latency.

Dense Wave Division Multiplexing (DWDM) promises to make huge amounts of bandwidth available inexpensively. By 1997 a single strand of optical fiber could carry 10 Gbps, with capacities of 400 Gbps and beyond expected in later years. Combined with Gigabit Ethernet technologies in enterprise networks, the notion of "free" bandwidth would presumably be the lure for networks to use existing data network technologies with fast Layer 2 and Layer 3 switches thereby obviating deployment of new ATM based technologies. Thus, "best effort" QoS with copious amounts of bandwidth is sufficient to serve the needs of a wide range of traffic types.

This, however, would suggest that waste be designed into networks as a matter of network design principle to ensure that networks not become congested. This strategy, which assumes that it is possible to keep up with, if not exceed bandwidth demand from users, may be adequate in enterprise networks, but is a dubious one for public networks. Furthermore, it assumes that "infinite" bandwidth will be deployed everywhere. Otherwise, there will be areas where network congestion does occur, thereby precluding real-time service over a unified multiservice network infrastructure in those cases.

In the integrated services Internet there will be real-time applications like packet voice and packet video which are UDP-based applications and do not implement the throttling mechanisms that are inherent in TCP. The requirements of real-time applications are such that there is no benefit in throttling traffic because a delayed packet would be of no use in an isochronous application. Thus, UDP-applications, which are unresponsive to network congestion, can effectively shut down TCP applications that do respond and throttle transmission under these circumstances. Consequently, it is important for the network to be able to rely upon mechanisms other than infinite bandwidth and best effort delivery to protect itself against network congestion.

In the integrated services Internet, even TCP is not a sufficiently reliable defense against congestion from the network provider perspective. The reason is that TCP maintains state information (for throttling, etc.) in the end station, not within the network. This creates a potential for abuse if aggressive TCP implementations are deployed. By an "aggressive" TCP implementation, we mean to refer to implementations which (relative to other TCP implementations) back off more slowly, and increase transmission window size more rapidly than less aggressive implementations. Vendors offering such implementations have the incentive of being able to market "faster TCP" software. Applications may also monopolize network bandwidth by opening multiple TCP connections to effect higher speed information transfer.

Network service providers using managed bandwidth are also able to pursue revenue opportunities through the deployment and trafficking of services based upon QoS. Thus, public network operators can offer their customers service level agreements (SLA) with usage-based billing according to QoS classes such as "Platinum," "Gold," or "Silver." As network bandwidth becomes greater, QoS-based SLAs create product differentiation and can help protect revenues by allowing network operators to avoid having bandwidth become a low-margin commodity item.

A key aspect of this strategy for networks based upon ATM requires that ATM technology be more than a core network technology. For a public carrier this means that the technology must find a presence in the customer premises environment. This may be accomplished through the deployment of broadband access multiplexors, which are able to take inputs from customer premises equipment that may not be based upon ATM:

- Circuit-switched voice and data from PBXs
- Data traffic from token ring and/or Ethernet LANs
- NTSC video from video codecs

The broadband access multiplexor is then able to provide interworking functions to put the various traffic inputs into ATM VCs to be sent across the UNI to the public ATM network.

2. Emerging Applications

2.1 Asymmetric Digital Subscriber Line (ADSL)

ADSL is a technology that allows high-speed access over twisted-pair cable. ADSL is asymmetric in that it behaves like different technologies in each direction between the subscriber and the central office. It provides a 144-kbps channel from the subscriber to the central office like an ISDN DSL, and a 1.536 Mbps channel from the central office to the subscriber, like a high bit rate DSL (HDSL).

ADSL service is being deployed by ILECs and data CLECs as a low cost means of providing standard telephony in addition to high-speed Internet access. A network architecture for ATM over ADSL consists of five elements, as shown in Figure 12-1.

1. Core ATM network—consisting of a network of one or more ATM switches.

Chapter 12: Future Directions for ATM

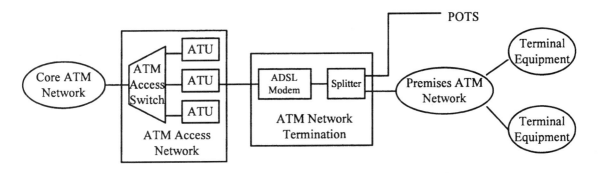

Figure 12-1. ATM over ADSL.

2. ATM Access Network—in the central office may consist of:
   - ATM access multiplexors, which perform ATM layer functions and concentrate traffic for access over an ATM network.
   - ADSL Transceiver Units (ATU), which support transmission over point-to-point copper media.
   - Passive splitters which separate voice and data channels.
   - A distribution network consisting of twisted-pair copper media.
3. Access Network Termination—at the customer premises, which may consist of:
   - ADSL Modems.
   - Passive splitters.
4. Customer Premises ATM Network.
5. Terminal Equipment.

2.2 Digital Television (DTV)

The Federal Communications Commission (FCC) mandated that television broadcasters in the United States begin changeover to DTV in the fall of 1998. DTV will require television networks to move massive amounts of digital data. The two DTV categories are Standard Definition Television (SDTV) and High Definition Television (HDTV). Raw SDTV requires a 270 Mbps data stream while HDTV requires 1.5 Gbps. Using Motion Picture Experts Group (MPEG-2) encoding brings these data rates down to 5 Mbps and 19.39 Mbps, respectively.

There are plans in the public carrier sector to offer SVC services over ATM networks using CBR and nrt-VBR to networks for the distribution of digital programming to affiliates. ATM allows bandwidth allocation on a per-connection basis. Thus, broadcasters using ATM realize benefits over use of the American National Standards Institute (ANSI) standard for video over broadband networks, which calls for DS-3 transport. Instead of requiring broadcasters to commission a partially used DS-3, they are able to obtain bandwidth on demand using SVC service.

The emergence of DTV also provides a basis for the emergence of the "virtual studio" where video postproduction is done independent of the location of the postproduction facilities over high-speed networks. This is an application that supports the deployment of ATM to the desktop to interconnect video editors in the virtual studio. This application requires high bandwidths: OC-3 to the desktop and OC-12 and higher rates in the enterprise network.

2.3 Wireless ATM

There are three areas of focus for ATM in Wireless networks that are discussed in this section:

1. Connection of Wireless Base Station Controllers (BSC)
2. Wireless ATM over Personal Communication Service (PCS)
3. Wireless ATM over Local/Multichannel Multipoint Distribution Service (LMDS/MMDS)

2.3.1 BSC Interconnection

In order to make best use of radio spectrum, cellular networks perform speech compression using vocoding to increase the number of calls that may be handled simultaneously by a given BSC. These calls are decompressed and carried over TDM facilities to a central office switch.

BSCs may be connected to an ATM switch by a SONET ring as shown in Figure 12-2. AAL Type 2 supports the variable nature of compressed speech transported between the BSC and the ATM switch. The ATM switch then aggregates traffic from the BSCs and performs interworking to the TDM network.

2.3.2 ATM Interworking with PCS

In this application the end users are PCS terminals that are connected via a PCS access network to a Mobility Enhanced (ME) ATM access switch as shown in Figure 12-3. The ME ATM access switch provides access to the ATM network and is responsible for performing IWF between the ATM network and the PCS network and for delivering the data stream to the proper BSC. This requires signalling between the ME ATM access switch and the PCS terminal to perform the appropriate handoffs between BSCs, and between ME ATM access switches, as necessary.

2.3.3 ATM over LMDS

LMDS is a wireless technology, which uses concepts similar to cellular telephone networks, that is suitable for broadband delivery of voice, data,

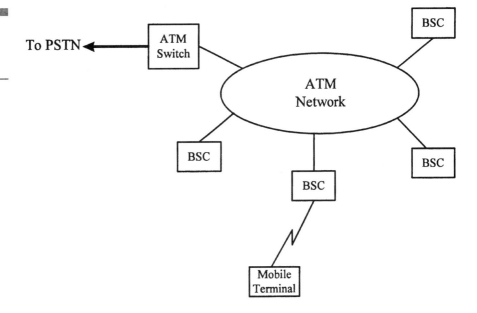

Figure 12-2. Interconnection of wireless BSCs via ATM.

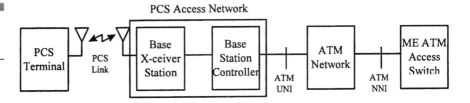

Figure 12-3. PCS interworking with ATM.

and video transmission. LMDS operates over the 28-GHz to 31-GHz frequency band and supports bit rates from 64 Kbps to 155 Mbps over a bidirectional wireless link. The FCC awarded licenses for portions of the LMDS spectrum in February 1998 with service deployment scheduled for late 1998 or early 1999. Since the network may be expanded one wireless link at a time, LMDS is very scalable.

Because ATM is designed to handle multiple types of data streams, it is being positioned by ATM equipment vendors as the perfect underlying medium for LMDS.

3. Future Challenges

Following the hyperbole that surrounded ATM initially came questions of "is ATM dead?" when ATM didn't live up to initial forecasts. ATM is far from dead, but legacy technologies are being adapted in response. RSVP is evolving as a means of providing end-to-end QoS over router-based IP networks. Gigabit Ethernet is being developed to support line rates in excess of 1 Gbps, with some implementation of link level QoS. IP over SONET provides the ability to move native IP traffic at high bit rates without ATM. And "voice over IP" is a development that, if proven viable in actual practice, significantly undercuts the need for ATM as the infrastructure for an integrated services network.

At this writing, much of the demand for ATM switching equipment is from public carriers. This includes interexchange carriers (IXC), incumbent local exchange carriers (ILEC), competitive LECs (CLEC), and increasingly Internet Services Providers (ISP) who see ATM, when used with the requisite edge devices and interworking functions, as providing the most effective platform for multiple services.

In wide area networks, ATM allows subscribers to purchase specific amounts of bandwidth on a per-VC basis rather than having to pay for unused bandwidth on a DS-3 or OC-3 interface. CBR and ATM circuit emulation service allow customers to retire private line T-1 circuits over wide area networks.

However, the ability to exploit the "bandwidth on demand" features of ATM requires the availability of SVC service, which is beginning to be offered by public carriers at the time of this writing. This allows data traffic over wide area networks to use UBR or ABR PVCs while other available bandwidth could be used on demand by SVC traffic. When demand for bandwidth from other sources is low, the unused bandwidth becomes available for use in transporting data until needed for other applications with more stringent service requirements. This capability allows great flexibility in the use of network bandwidth.

For all of the bandwidth management and QoS capabilities of ATM, there is a price, and that price can be paid in a loss of efficient use of bandwidth. As we have seen, under some circumstances ATM can be an inefficient means of carrying data traffic relative to IP over SONET. However, we have also seen that ATM can be more efficient at carrying voice, and with less end-to-end delay. Furthermore, the small, fixed length ATM cell also reduces end-to-end jitter when carrying real-time traffic relative to variable length packet-based networks.

Ostensibly this might suggest that the only possible application for ATM is in public networks where management of huge amounts of bandwidth and the constraints associated with SLAs are major concerns. Where these considerations may not be as significant is in enterprise networks. Thus, enterprise networks would, therefore, connect into public carrier networks usings IP over SONET because of its greater efficiency when carrying data traffic—ATM core networks interconnecting to a *non*-ATM subscriber world.

But the subscriber picture for ATM interconnection is actually a much more efficient friendlier one than it would first appear. Subscribers may connect to public carrier networks offering a Frame Based UNI (FUNI) using guaranteed frame rate (GFR) service and realize greater efficiencies than can be realized with cell-based ATM. FUNI allows up to 9,232 octets of payload to be carried with as little as 6 octets of overhead at rates of at least DS1/E1.

Another challenge facing ATM is in reducing the high people-related costs of ATM deployment. The training necessary for existing staff to develop expertise with ATM technology can be a multiyear undertaking. Thus, unless there is a strongly compelling reason for deployment of ATM, the commitment required by ATM creates a disincentive to all but the early adopters. The colleges and universities play a major role in developing the basis for proliferating ATM expertise within industry by providing instruction and training to its students.

The risks of putting more services and more critical services on a single network produce greater ramifications in the event of a failure in the network. Deployment of new technologies is always an unpredictable process. This was true of ATM, which in its early days was plagued by divergent standards, proprietary implementations, and interoperability problems among different vendors' problems. Emerging "hot" technologies face the same problems, so ATM has a head start. Still, uncertainties about the ability of ATM to actually deliver on the promise of providing an integrated multiservice network dictate a "go slow" approach toward wholesale replacement of existing networks, especially the proven and reliable PSTN, for which there are billions of dollars of invested infrastructure, and which adequately handles voice telephony services. Thus, ATM is more likely to be an evolutionary, rather than revolutionary communications technology.

ACRONYMS

A

ANSI American National Standards Institute
ADSL Asymmetric Digital Subscriber Line
ATM Asynchronous Transfer Mode
AAL ATM Adaptation Layer
ATMARP ATM Address Resolution Protocol
ABR Available Bit Rate

B

B-ICI Broadband Inter-Carrier Interface
B-ISDN Broadband Integrated Services Digital Network
B-ISUP Broadband ISDN User Part (B-ISUP)
BUS Broadcast and Unknown Server

C

CDV Cell Delay Variation
CDVT Cell Delay Variation Tolerance.
CER Cell Error Ratio
CLP Cell Loss Priority (CLP)
CLR Cell Loss Ratio (CLR)
CMR Cell Misinsertion Rate (CMR)
CTD Cell Transfer Delay (CTD)
CAS Channel Associated Signalling
CSU Channel Service Unit
CES Circuit Emulation Service
CPCS Common Part Convergence Sublayer
CN Congestion Notification
CONS Connection Oriented Network Service
CLNS Connectionless Network Service

CBR Constant Bit Rate (CBR)
CLS Controlled Load Service (CLS)
CS Convergence Sublayer (CS)
CAC Connection Admission Control (CAC)
CPE Customer Premises Equipment

D

DCE Data Circuit-Terminating Equipment (DCE)
DLL Data Link Layer (DLL)
DTE Data Terminal Equipment (DTE)
DSU Digital Service Unit (DSU)
DSn Digital Signal Level-n

E

EPD Early Packet Discard
EMS Element Management System
EFCI Explicit Forward Congestion Indication

F

FDDI Fiber Distributed Data Interface
FA Frame Address
FID Frame Identification
FUNI Frame User to Network Interface

G

GCRA Generic Cell Rate Algorithm
GFC Generic Flow Control (GFC)
GFR Guaranteed Frame Rate
GS Guaranteed Service

H

HEC Header Error Control
HIPPI High Performance Parallel Interface

I

IEEE Institute of Electrical and Electronics Engineers
ILMI Integrated Local Management Interface
ISDN Integrated Services Digital Network
ITU International Telecommunication Union
IETF Internet Engineering Task Force (IETF)
IIS Internet Integrated Service Model
IP Internet Protocol (IP)
IPv4 IP version 4
IPv6 IP version 6
IWF Interworking Function

L

LTE Line Terminating Equipment
LAN Local Area Network
LANE LAN Emulation
LE ARP LAN Emulation Address Resolution Protocol
LEC LAN Emulation Client
LECS LANE Emulation Configuration Server
LES LANE Emulation Server
LIS Logical IP Subnet
LLC Logical Link Control

M

MIB Management Information Base
MBS Maximum Burst Size
MAC Media Access Control
MCR Minimum Cell Rate
MARS Multicast Address Resolution Server
MCS Multicast Server
MAPOS Multiple Access Protocol over SONET/SDH
MPOA Multi-Protocol over ATM
MPC MPOA Client
MPS MPOA Server (MPS)

N

NMS Network Management System
NNI Network Node Interface
NPC Network Parameter Control
NSAP Network Service Access Point
NHRP Next Hop Resolution Protocol
NBMA Non-Broadcast Multiple Access

O

OAM&P Operations, Administration, Maintenance, and Provisioning

P

PPD Partial Packet Discard
PTE Path Terminating Equipment
PTI Payload Type Identifier
PCR Peak Cell Rate
PVC Permanent Virtual Connection
PMD Physical Medium Dependent
PDH Plesiochronous Digital Hierarchy
PPP Point to Point Protocol
PNNI Private Network Node Interface
PCI Protocol Control Information
PDU Protocol Data Unit
PRI ISDN Primary Rate Interface
PRM Protocol Reference Model

Q

QoS Quality of Service

R

RED Random Early Drop
RTP Real Time Protocol

RTCP RTP Control Protocol
RFC Request For Comments
RM Resource Management Cell
RSVP Resource Reservation Protocol

S

STE Section Terminating Equipment
SAR Segmentation and Reassembly
SAP Service Access Point
SDU Service Data Unit
SSCS Service Specific Convergence Sublayer
SSCOP Service Specific Connection Oriented Protocol
SSCF Service Specific Coordination Function
SECBR Severely Errored Cell Block Ratio
SAAL Signalling AAL
SEAL Simple and Efficient AAL
SNMP Simple Network Management Protocol
SVPC Soft PVC
SNAP Subnetwork Attachment Point
SCR Sustained Cell Rate
SMDS Switched Multi-megabit Data Service
SVC Switched Virtual Connection
SAC Synchronous to Asynchronous Conversion
STM Synchronous Transfer Mode
SDH Synchronous Digital Hierarchy
SONET Synchronous Optical Network

T

TMN Telecommunications Management Network
TDM Time Division Multiplexing
TCP Transmission Control Protocol
TC Transmission Convergence Sublayer

U

UBR Unspecified Bit Rate
UPC Usage Parameter Control
UDP User Datagram Protocol
UNI User-Network Interface

V

rt-VBR Variable Bit Rate-Real Time
nrt-VBR Variable Bit Rate-Non Real Time
VC Virtual Channel, Virtual Circuit
VCC Virtual Channel Connection
VCI Virtual Channel Identifier
VCL Virtual Channel Link
VLAN Virtual LAN
VP Virtual Path
VPC Virtual Path Connection
VPCI Virtual Path Connection Identifier
VPI Virtual Path Identifier
VPL Virtual Path Link
VPT Virtual Path Terminator

W

WWFQ Weighted Fair Queuing
WRED Weighted Random Early Drop

GLOSSARY

A

AAL CS. Provides an interface between the AAL and higher layer protocols. The AAL CS is divided into two components: a Common Part Convergence Sublayer (CPCS) and a Service Specific Convergence Sublayer (SSCS).

American National Standards Institute (ANSI). An organization that approves national standards in the United States and develops positions for the United States in international standards organizations.

Asymmetric Digital Subscriber Line (ADSL). A technology that delivers high bandwidth over conventional copper wire at limited distances. ADSL is designed to deliver higher data rates across a UNI from the network to customer direction.

Asynchronous Transfer Mode (ATM). A transfer mode in which information is organized into 53 octet blocks known as cells. ATM is asynchronous in the sense that network resources are only used when the traffic source is actually transmitting cells. Compare with Synchronous Transfer Mode.

ATM Adaptation Layer (AAL). A collection of standardized protocols that adapt traffic from higher protocol layers to the ATM cell format. The AAL is subdivided into two basic sublayers: Convergence Sublayer (CS) and Segmentation and Reassembly (SAR).

ATM Address Resolution Protocol (ATMARP). A mechanism for translating a non-ATM address into a corresponding ATM address for routing across an ATM network.

ATM Forum. An industry consortium that issues implementation agreements pertaining to ATM systems.

ATM Layer. The PRM layer responsible for the transparent transfer of fixed size ATM cells between communicating higher layer entities.

Available Bit Rate (ABR). An ATM service category that provides low cell loss delivery for traffic at rates up to a minimum cell rate and best effort delivery for higher traffic rates (compare with UBR).

B

Bandwidth. (1) The range of frequencies that can be transmitted over a communications channel or transmission medium. (2) The amount of data that can be sent through a given communications circuit per unit of time.

Bearer Channel. A VC that carriers user information between communicating endpoints.

Bearer Service. A definition of the underlying characteristics of a VC that provides the capability for the transfer of information between communicating endpoints.

Best Effort Delivery. Describes the services provided by a VC with no specified values for any of the QoS parameters.

Broadband. Transmission rates which exceed ISDN primary rate.

Broadband Integrated Services Digital Network (B-ISDN). An ITU communication standard which supports the integrated, high-speed transmission of data, audio, and video over a single network infrastructure.

Broadband Inter-Carrier Interface (B-ICI). (1) An NNI that defines the interconnection point between ATM switches that are within different public carrier B-ISDN networks. (2) An implementation agreement by the ATM Forum which defines signalling and internetworking procedures across this interface.

Broadband ISDN User Part (B-ISUP). An ITU recommendation which defines signalling procedures over Signalling System Number 7 (SS7) networks.

Broadcast. The sending of a data packet to all nodes on a network simultaneously.

Broadcast and Unknown Server (BUS). A LANE entity that provides broadcast and multicast services. In addition, the BUS forwards data to destination LECs whose ATM addresses are unknown to the source LECs.

Burstiness. A measure of the variation in the rate of traffic from a source that is defined as the ratio of PCR to average cell rate.

C

Cell. An ATM Layer PDU. The basic 53-octet unit of information transfer in an ATM network. Each unit consists of a 5-octet header and a 48-octet payload.

Cell Delay Variation (CDV). The variation in CTD among cells in a connection. A maximum (or peak) CDV value is a QoS parameter.

Cell Delay Variation Tolerance (CDVT). A parameter that defines the acceptable amount of burstiness in cell traffic submitted across a UNI or B-ICI.

Cell Error Ratio (CER). A QoS parameter that measures the ratio of errored cells to the total number of cells transmitted. A cell is "errored" if it contains an uncorrectable error in the cell header.

Cell Loss Priority (CLP). A field within the ATM cell header used to distinguish cells with high versus low delivery priority within an ATM network. Cells with low CLP may be discarded first during periods of network congestion.

Cell Loss Ratio (CLR). A QoS parameter that measures the ratio of cells lost to the total number of cells transmitted. A cell is declared "lost" if it does not arrive within a maximum CTD.

Cell Misinsertion Rate (CMR). A QoS parameter that measures the number of misinserted cells over a measurement time interval. A misinserted cell is one that is received at a connection endpoint that was not transmitted by the peer endpoint in the connection.

Cell Relay. A network technology based on the transfer of information in small, fixed length packets, or cells.

Cell Transfer Delay (CTD). A QoS parameter that measures the time it takes to transfer an ATM cell from a source to its destination over a VC.

Channel. A transmission path between two communicating endpoints.

Channel Associated Signalling (CAS). A method of signalling in which signalling information is carried over the bearer channel. Also known as "robbed bit" signalling.

Channel Service Unit (CSU). Digital interface device that connects end-user equipment to the local digital telephone loop.

Circuit Emulation Service (CES). A capability that allows an ATM network to carry traffic from a TDM network.

Circuit Switching. A switching technology in which network resources are dedicated to a specific communication path upon establishment and remain dedicated until the communication path is terminated. Compare with Packet Switching.

Classical IP over ATM. A protocol that allows IP traffic to be carried over ATM networks but requires that routes crossing an LIS boundary use an intermediate router.

Common Part Convergence Sublayer (CPCS). A component of the AAL CS that performs error checking and other functions that are independent of higher layer protocols. The CPCS provides an interface between the SAR and SSCS sublayers.

Congestion Notification (CN). A field in a FUNI frame header that carries information equivalent to the EFCI subfield within the PTI field of an ATM cell header.

Connection. A concatenation of ATM Layer links that provides an end-to-end information transfer capability between communicating endpoints.

Connection Admission Control (CAC). The set of actions taken during SVC establishment to determine whether a connection request will be accepted or rejected.

Connectionless Network Service (CLNS). A network service which does not require the establishment of a fixed communication path prior to the beginning of communication between the endpoints.

Connection Oriented Network Service (CONS). A network service in which a communication between endpoints occurs over a fixed communication path which is established prior to the beginning of communication between the endpoints.

Constant Bit Rate (CBR). An ATM service category that supports the transmission of continuous bit rate traffic. The QoS guarantees associated with CBR makes it suitable for isochronous traffic such as voice, video, and other real-time services.

Controlled Load Service (CLS). An IIS model service that does not specify an upper bound on end-to-end delay for traffic carried across the network.

Convergence Sublayer (CS). Provides an interface for PDUs passed between a protocol sublayer and the next higher protocol sublayer. Convergence sublayers within ATM include the Transmission Convergence Sublayer (TC), Common Part Convergence Sublayer (CPCS), and Service Specific Convergence Sublayer (SSCS).

Customer Premises Equipment (CPE). Communications terminating equipment installed at customer sites and connected to a public carrier communications network.

Cut-Through Route. A route across an ATM network which crosses LIS boundaries without using an intermediate router.

D

Data Circuit-Terminating Equipment (DCE). Device at the network end of a UNI that forwards traffic, and provides a clocking signal used to synchronize data transmission between DCE and DTE devices.

Datagram. A logical grouping of information sent from the Network Layer over a transmission medium without prior establishment of a connection between source and destination.

Data Link Layer (DLL). The layer of the PRM that is responsible for data transfer across a single physical connection, or series of physical connections, between Network Layer entities.

Data Terminal Equipment (DTE). Device at the user end of a UNI that serves as a data source, destination, or both.

Dead Cells. Cells that carry packet information that will be discarded at the destination because a cell segment within the packet has been lost.

Digital Section. A transmission medium, together with associated equipment, required to provide a means of transporting information between network elements that have the capability of multiplexing or demultiplexing traffic.

Digital Service Unit (DSU). Device used in digital transmission that adapts basic rate and subrate interfaces on DTE to primary rate transmission facilities.

Digital Signal (DS) Level. One of several transmission rates in the Plesiochronous Digital Hierarchy.

Dual Homing. A network topology in which a device is connected to a network by way of two independent attachment points in which one attachment point is the primary connection and the second is a standby connection used in the event of a failure of the primary connection.

Dynamic Feedback Mechanism. Procedures that support the exchange of information between the network and end systems to dynamically regulate the allowable rate of information flow on a connection.

E

Early Packet Discard (EPD). A technique for discarding all cells that comprise a packet based upon a prediction of whether the packet can be successfully delivered to the destination without cell loss.

Edge Device. A switching element which implements IWF capabilities.

Element Management System (EMS). A system responsible for setting the hardware and software parameters used in managing a single network element.

Encapsulation. The wrapping of data in a particular PDU structure with a header that identifies the contents to a receiving protocol entity.

Ethernet. A LAN specification developed by Xerox Corporation. Similar to IEEE 802.3 series of standards.

Explicit Forward Congestion Indication (EFCI). A subfield within the PTI that indicates network congestion.

F

Fiber Distributed Data Interface (FDDI). A LAN standard defined by ANSI X3T9.5 that specifies a 100 Mb/s network using fiber-optic cable, with transmission distances of up to 2 km.

Flowspec. A traffic contract definition within RSVP.

Frame. A logical grouping of information with a specific format that includes specific bit patterns to delineate the beginning and/or end.

Frame Address (FA). The routing field in a FUNI frame header. Subfields within this field map to VPI and VCI values in an ATM cell header.

Frame Identification (FID). A field in a FUNI frame header which carries information that is a subset of that which is carried in the PTI field of an ATM cell header.

Frame Relay. A network technology based on the transfer of information in variable length packets or frames.

Frame User to Network Interface (FUNI). A UNI that provides ATM services to frame-based traffic.

FRF.5 (Frame Relay/ATM PVC Network Interworking Implementation Agreement). An implementation agreement defined by the Frame Relay Forum that specifies the tunneling of Frame Relay frames over ATM networks.

FRF.8 (Frame Relay/ATM PVC Service Interworking Implementation Agreement). An implementation agreement defined by the Frame Relay Forum that specifies the interworking of Frame Relay service and ATM Cell Relay service.

G

G.711. An ITU recommendation which uses 64 kb/s analog to digital conversion of voice frequency analog signals.

G.728. An ITU recommendation which uses 16 kb/s analog to digital conversion of voice frequency analog signals.

G.729. An ITU recommendation which uses 8 kb/s analog to digital conversion of voice frequency analog signals.

Generic Cell Rate Algorithm (GCRA). A conceptual model for determining compliance to the traffic contract for submitted traffic over a connection from a traffic source.

Generic Flow Control (GFC). A control field within the ATM cell header that can be used to regulate the allowable rate of information flow across a UNI.

Guaranteed Frame Rate (GFR) Service. An ATM Forum defined service definition that defines a traffic contract in terms of frames rather than cells.

Guaranteed Service (GS). An IIS model service that specifies an upper bound on end-to-end delay for traffic carried across the network.

H

H.323. An ITU recommendation that defines a set of procedures for multimedia communications services over packet based networks. The emerging standard for voice over the Internet.

Header. The 5-octet data portion of a 53-octet ATM cell. Also used to describe PCI placed before the PDU portion of an SDU (compare with Trailer).

Header Error Control (HEC). A field within the ATM cell header used by the Physical Layer to detect errors in the ATM cell header and to delineate individual cells.

Heterogeneous Branch Point. A multicast distribution point defined within RSVP in which the flowspecs of the individual branches are not the same.

High Performance Parallel Interface (HIPPI). A high performance interface standard defined by ANSI. HIPPI is typically used to connect supercomputers to peripherals and other devices.

I

Idle Cell. An unassigned cell.

IEEE 802.3. A LAN protocol defined by the IEEE that is similar to Ethernet.

Institute of Electrical and Electronics Engineers (IEEE). A professional organization whose activities include the development of communications and network standards.

Integrated Local Management Interface (ILMI). An ATM Forum implementation agreement which specifies the use of SNMP and an ATM MIB to provide network status and configuration information across a UNI or Private NNI.

Integrated Services Digital Network (ISDN). A network technology that delivers voice and digital network services over a single loop.

Interface. (1) A connection between two systems or devices that defines physical interconnection characteristics, signal characteristics, and semantics of interchanged signals. (2) A boundary between protocol sublayers contained within a PRM layer.

International Telecommunication Union (ITU). An international body which, under the auspices of the United Nations, establishes international telecommunications standards.

Internet Engineering Task Force (IETF). A task force responsible for developing Internet standards.

Internet Integrated Service (IIS) Model. A proposed enhancement of the Internet architecture to support both data and real-time services that includes traffic management concepts that are similar to those of ATM.

Internet Protocol (IP). A Network Layer protocol which offers connectionless delivery of datagrams.

Interworking Function (IWF). A capability that allows an ATM network to carry traffic that originates or terminates on a non-ATM network.

IP Subnet. A subcomponent of a network structured (with other subnets) to provide a multilevel, hierarchical routing structure.

IP version 4 (IPv4). The version of IP as specified in IETF RFC 791.

IP version 6 (IPv6). The replacement for IPv4 that includes a flow identifier field in the packet header that can be used to label end-to-end packet flows across an IP network.

ISDN Primary Rate Interface (PRI). An ISDN UNI that consists of a frame of 24 or 32 64 Kb/s channels.

Isochronous. Signals that are dependent on externally supplied uniform timing or carry their own timing information embedded as part of the signal.

L

Latency. The delay between the time a traffic source sends a data frame and the time that the frame is received at the destination. Also referred to as End-to-End Delay.

Layer 3 Switching. The routing of Network Layer packets based upon stored information instead of performing route computation upon arrival of each packet.

Line Terminating Equipment (LTE). Equipment which is located at the endpoints of a digital section.

Local Area Network (LAN). A high-speed network that interconnects workstations over short distances.

LANE Emulation Configuration Server (LECS). A LANE entity that is responsible for assigning LECs to the appropriate LESs.

LANE Emulation Server (LES). A LANE entity that is responsible for responding to LE ARP requests from LECs.

LAN Emulation Address Resolution Protocol (LE ARP). A LANE protocol for translating between LAN addresses and corresponding ATM addresses.

LAN Emulation Client (LEC). A LANE entity that performs data forwarding, address resolution, and other LANE related control functions.

LAN Emulation (LANE). A protocol which allows an ATM network to give the appearance of a LAN by emulating features and the MAC layer software interface of traditional LANs.

LLC/SNAP Encapsulation. A method for encapsulating information from higher layer protocols for transport over an ATM network using a single VC.

Logical IP Subnet (LIS). A group of IP nodes that connect to a single ATM network and belong to the same IP subnet.

Logical Link Control (LLC). A protocol developed by the IEEE for data-link layer transmission control.

Loop. A communication path between a subscriber and communication switching equipment in a network.

M

Managed Object. An abstraction representing the properties of a network resource as viewed from an NMS.

Management Information Base (MIB). A collection of managed objects.

Maximum Burst Size (MBS). The maximum number of consecutive cells that may be submitted at the PCR for an ATM connection using this parameter.

Media Access Control (MAC). The lower sublayers of the Data Link Layer that handles access to shared media networks.

Meta-Signalling. The capability for the establishment of SVCs between communicating endpoints with multiple VCCs assigned at each UNI. Meta-signalling allows multiple signalling endpoints to share a single UNI.

Minimum Cell Rate (MCR). The minimum cell transmission rate that will always be guaranteed over an ATM connection using this parameter.

Multicast. The capability of sending a data packet simultaneously to a specific subset of nodes in a network.

Multicast Address Resolution Server (MARS). A function that is similar to ATMARP that translates an IP multicast address to a corresponding list of ATM addresses.

Multicast Server (MCS). An ATM endpoint which maintains a point to multipoint connection with each ATM endpoint in a multicast group.

Multiple Access Protocol over SONET/SDH (MAPOS). A method of encapsulating IP datagrams in a SONET payload.

Multi-Protocol over ATM (MPOA). A protocol that is, in principle, capable of carrying multiple Network Layer protocols over ATM networks. MPOA provides the capability to interconnect LANs emulated by LANE.

MPOA Client (MPC). An MPOA virtual routing entity which performs packet forwarding.

MPOA Server (MPS). An MPOA virtual routing entity which performs route computation.

N

Narrowband. Transmission rates which are at or below ISDN primary rate.

Network Management System (NMS). A system responsible for setting the hardware and software parameters used in managing a network consisting of one or more network elements.

Network Node Interface (NNI). (1) An interface between ATM switches within a private network. (2) An interface between ATM switches within a public network.

Network Parameter Control (NPC). The set of actions used to enforce compliance with the traffic contract for traffic submitted across a B-ICI.

Network Resource. Any physical or logical entity whose behavior influences communications over a network.

Network Service Access Point (NSAP). The SAP between the Network and Transport Layers.

Next Hop Resolution Protocol (NHRP). A protocol that allows IP traffic to be carried over ATM networks which provides support for cut-through routes on NMBA networks.

Non-Broadcast Multiple Access (NBMA) Network. A term describing a network which allows multiple devices to connect and communicate simultaneously but does not support broadcasting.

Null AAL. An AAL type in which ATM cells are passed directly to a higher layer protocol entity with no processing by the AAL. Also referred to as AAL Type 0.

O

OAM Cell. An ATM cell which carries OAM&P information.

Octet. An 8-bit grouping of information.

One Armed Router. A router which receives and forwards traffic over a single ATM interface.

Operations, Administration, Maintenance, and Provisioning (OAM&P). The set of tasks which provide for overall management of a communications network.

P

Packet. A logical grouping of information that contains a header and/or trailer, and a payload.

Packet Forwarding. The transfer of a packet from one node to the next node along an end-to-end route between the source node and destination node.

Packet Switching. A switching technology in which network resources are not dedicated to a specific communication path. Network resources are used when there is information to be transferred over the communication path and are available for use by other communication paths at other times. Compare with Circuit Switching.

Partial Packet Discard (PPD). A technique for reducing the amount of resources on a ATM network that are wasted in the transport of dead cells.

Path Terminating Equipment (PTE). Equipment that is located at the ends of a transmission path.

Payload. (1) The portion of a cell, frame, or packet that contains upper layer information. (2) The 48-octet data portion of a 53-octet ATM cell.

Payload Type Identifier (PTI). A field within the ATM cell header that provides information that indicates how a received ATM cell is to be processed.

Peak Cell Rate (PCR). The maximum rate at which a traffic source may transmit cells over a connection.

Permanent Virtual Connection (PVC). A virtual connection which is established and terminated by manual procedures.

Physical Connection. A communication path between communicating endpoints in a circuit-switched network. Compare with Virtual Connection.

Physical Layer. The bottom layer in the PRM. It consists of a Transmission Convergence (TC) sublayer, and a Physical Medium Dependent (PMD) sublayer. It provides ATM cell transmission over physical interfaces and media that connect ATM devices.

Physical Medium Dependent (PMD). A sublayer within the Physical Layer that specifies bit timing and other characteristics specific to a particular transmission medium. Compare with Transmission Convergence sublayer.

Plesiochronous Digital Hierarchy (PDH). A collection of discrete transmission rates which may be sent over a network from traffic sources which are not synchronized to the same network clock.

Point-to-Multipoint Connection. A connection between a communicating endpoint, known as a root node, and a set of two or more communicating endpoints, known as leaf nodes, which provides a capability for the root node to perform broadcast communication with the leaf nodes.

Point-to-Point Connection. A connection between a pair of communicating endpoints.

Point-to-Point Protocol (PPP). A method for managing packet traffic sent over serial point-to-point links.

Primitive. An abstract interaction between PRM layers across an SAP.

Private Network. A network whose resources are dedicated to the transport of traffic from a single organizational entity.

Private Network Node Interface (PNNI). (1) An interface between ATM switches within a private ATM network. (2) A protocol that specifies signalling and routing within private ATM networks.

Protocol. A formal set of rules and conventions that defines the interactions between protocol layer entities that govern how devices work on a network exchange information.

Protocol Control Information (PCI). Contains information used within a PRM protocol layer to process a PDU received from a lower protocol layer. Also known as a header.

Protocol Data Unit (PDU). The basic unit of information upon which PRM protocol layer operates. An SDU received from a higher protocol layer becomes the PDU when used within that lower protocol layer.

Protocol Entity. An instance of a given protocol or protocol layer.

Protocol Reference Model (PRM). A model which describes the relationship between layers of a protocol.

Public Network. A network whose resources are used to transport traffic from a (generally unrestricted) set of individuals and organizational entities.

Q

Q.2931. An ITU recommendation which defines signalling procedures across an UNI. Also known as Digital Subscriber Signalling System Number 2 (DSS 2).

Quality of Service (QoS). Refers to the set of ATM performance parameters that characterizes the traffic over a given virtual connection. These parameters include maximum CTD, CDV, CLR, CER, SECBR, and CMR.

R

Random Early Drop (RED). A technique that assigns a probability of packet drop to packets arriving at an output queue.

Real Time Protocol (RTP). A transport protocol which supports real-time applications.

Regenerator Section. A transmission medium, together with associated equipment, required to regenerate a signal on the medium.

Request for Comments (RFC). A document series issued by the IETF that is the primary means for communicating information about the Internet.

Resource Management (RM) Cell. An ATM cell that carries information over an end-to-end connection needed to support the dynamic feedback mechanism of the ABR service category.

Resource Reservation Protocol (RSVP). A protocol that supports the IIS model by allowing for the reservation of resources on IP networks for carriage of packet traffic.

Route Computation. The processing of a routing protocol to determine a route for sending traffic from a source to a destination.

Routing. The process of finding a path between a source and a destination within a network. This process consists of route computation, and packet forwarding functions.

RTP Control Protocol (RTCP). A companion protocol to RTP that provides feedback information on the QoS being provided to applications over an RTP media channel.

RTP Media Channel. A channel used within the H.323 specification for the transport of audio and video information streams.

S

Section Terminating Equipment (STE). Equipment that is located at the ends of a regenerator section.

Segmentation and Reassembly (SAR). An AAL sublayer that segments variable length PDUs received from the CPCS into fixed length ATM cells.

Service Access Point (SAP). A point at which a PRM layer presents its services to the next higher protocol layer.

Service Data Unit (SDU). A logical grouping of information containing a PCI and a PDU. Within the OSI protocol reference model the SDU is the unit of information passed from a protocol layer to the next lower protocol layer.

Service Interworking. The mapping of services from a non-ATM network to equivalent services within an ATM network.

Service Specific Connection Oriented Protocol (SSCOP). A component of the SAAL which provides reliable transfer of signalling information between signalling endpoints. SSCOP provides an interface between the CPCS and SSCF.

Service Specific Convergence Sublayer (SSCS). A component of the AAL CS that performs functions that are dependent on higher layer protocols. The SSCS allows interworking of ATM with other services (such as Frame Relay) by adapting the services of the higher layer protocol to an equivalent ATM representation. The SSCS provides an interface between the CPCS and the higher layer protocol.

Service Specific Coordination Function (SSCF). A component of the SAAL that maps SSCOP services to higher layer protocols.

Severely Errored Cell Block Ratio (SECBR). A QoS parameter that measures the ratio of contiguous blocks of transmitted ATM cells with an unacceptably high number of errored cells to the total number of transmitted blocks of ATM cells.

Signal. An abstract interaction between protocol sublayers within a PRM layer.

Signalling. The capability for the establishment of SVCs between communicating endpoints over a single VCC for each UNI.

Signalling AAL (SAAL). AAL Type 5 with an SSCS which interfaces with higher layer signalling protocols. The SAAL SSCS consists of a Service Specific Connection Oriented Protocol (SSCOP) sublayer and a Service Specific Coordination Function (SSCF) sublayer.

Signalling Channel. A VC that is separate from the bearer channel that carries signalling information between communicating endpoints.

Simple and Efficient AAL (SEAL). Currently known as AAL Type 5.

Simple Network Management Protocol (SNMP). A protocol that provides a means to monitor and control network devices.

Soft PVC (SPVC). A PVC which is initiated by a request from an EMS but is established across the network using signalling procedures.

Sublayer. A logical subdivision of a protocol layer.

Subnetwork Attachment Point (SNAP). (1) Identifies the higher layer protocol when used with LLC/SNAP encapsulation. (2) An address that represents a unique address used by a station attached to a particular subnetwork.

Sustained Cell Rate (SCR). The rate at which a bursty traffic source may transmit cells over a time scale that is long compared to the time scale for which cells may be transmitted at the PCR for an ATM connection using this parameter.

Switched Multi-megabit Data Service (SMDS). A high speed, connectionless data transport service defined by Bellcore that provides for the transfer of variable length packets.

Switched Virtual Connection (SVC). A virtual connection which is established and terminated on-demand through signalling procedures between the communicating endpoints.

Switching. The process of connecting appropriate transmission facilities to form a desired communication path between two communicating endpoints.

Synchronous Digital Hierarchy (SDH). An international standard that defines a collection of discrete transmission rates which may be sent over a network from traffic sources which are synchronized to the same network clock.

Synchronous Optical Network (SONET). A network specification developed by Bellcore that defines a collection of discrete transmission rates that may be sent over optical fiber from traffic sources that are synchronized to the same network clock.

Synchronous to Asynchronous Conversion (SAC). An IWF capability that performs a translation between STM frames and ATM cells.

Synchronous Transfer Mode (STM). A transfer mode in which information from a traffic source is assigned to specific time slots in the network that are reserved for that traffic source whether or not the source is actually transmitting information organized into 53 octet blocks known as cells. See also Time Division Multiplexing. Compare with Asynchronous Transfer Mode.

T

Time Division Multiplexing (TDM). A method of serving a number of simultaneous channels over a common transmission path by assigning the transmission path sequentially to the various channels, each assignment being for a discrete time interval.

Time slot. A fundamental unit of bandwidth in a TDM network.

Traffic Contract. The specification of a traffic descriptor and the QoS parameters which will be provided to traffic from a source which is compliant with that traffic descriptor.

Traffic Descriptor. The set of parameters that characterize a traffic source.

Traffic Policing. UPC or NPC.

Traffic Shaping. Techniques used to alter the cell transfer rate to bring a traffic source in compliance with a traffic contract.

Trailer. Describes PCI placed after the PDU portion of an SDU (compare with Header).

Transfer Mode. A set of methods covering transmission, multiplexing, and switching in a telecommunications environment.

Transmission. The process of sending information from one point to another in a communications network over some medium.

Transmission Control Protocol (TCP). A connection-oriented transport layer protocol that provides reliable full-duplex data transmission.

Transmission Convergence Sublayer (TC). Performing physical layer protocol functions which are not dependent upon a specific transmission medium. Provides an interface between the Physical Layer and the ATM layer. See also Physical Medium Dependent (PMD) sublayer.

Transmission Path. A logical connection between the point at which information is assembled into a standard frame format for transport over a transmission medium, and the point at which the frame format is disassembled.

Trunk. A communication path between communication switching equipment in a network.

Tunneling. The encapsulation of a higher layer protocol for transport over an ATM network such that, to the higher layer protocol, the ATM network appears as a single hop irrespective of the actual number of hops transited through the ATM network.

U

Unassigned Cell. A cell that does not contain information from an application using the ATM Layer service.

UNI version 3.x and UNI version 4.0. A series of implementation agreements by the ATM Forum which defines signalling and information transfer procedures across a UNI.

Unspecified Bit Rate (UBR). An ATM service category that provides best effort delivery.

Usage Parameter Control (UPC). The set of actions used to monitor and control traffic submitted by end users across an UNI.

User Datagram Protocol (UDP). A connectionless transport layer protocol that exchanges datagrams without providing mechanisms for acknowledgments or guaranteed delivery.

User-Network Interface (UNI). (1) An interface point between ATM end users and an ATM switch. (2) An interface between a private network ATM switch and a public network ATM switch.

V

Variable Bit Rate-Non-Real Time (nrt-VBR). An ATM service category that differs from rt-VBR in that it does not provide QoS guarantees on maximum CTD and CDV.

Variable Bit Rate-Real Time (rt-VBR). An ATM service category that provides QoS guarantees on maximum CTD and CDV which makes it suitable for carrying variable bit rate traffic which have real time constraints, such as compressed video and audio.

VC Based Multiplexing. A method for encapsulating information from higher layer protocols for transport over an ATM network in which traffic for each protocol is carried over a separate VC.

VC Switch. An ATM network element that performs VCI and VPI translation and provides signalling functionality.

Virtual Channel (VC). A communication path that provides for the sequential unidirectional transport of ATM cells.

Virtual Channel Connection (VCC). A unidirectional concatenation of VCLs that connects peer AAL protocol entities that form the VCC endpoints.

Virtual Channel Identifier (VCI). A routing field within the ATM cell header that identifies a specific VCC.

Virtual Channel Link (VCL). A unidirectional path between the point where a VCI value is assigned to the point where it is translated or removed.

Virtual Channel Switch. A network element which connects VCLs.

Virtual Connection. A communication path between communicating endpoints in a packet switched network. Compare with Physical Connection.

Virtual LAN (VLAN). A set of workstations that are configured so that they can communicate as though they were attached to the same LAN even though they may be physically connected to different LAN segments.

Virtual Path (VP). A term used to describe the transmission path of a bundle of VCs that are transported as a group transparently over an ATM network.

Virtual Path Connection (VPC). A grouping of VCCs that are carried transparently between VCL endpoints.

Virtual Path Connection Identifier (VPCI). An identifier that is associated with a VPC whose value is constant across VPLs.

Virtual Path Identifier (VPI). A routing field within the ATM cell header that identifies a specific VPC.

Virtual Path Link (VPL). A unidirectional path between the point where a VPI value is assigned to the point where it is translated or removed.

Virtual Path Switch. A network element which connects VPLs.

Virtual Path Terminator (VPT). The endpoint of a VPC. Same as a VCL endpoint.

Virtual Routing. The division of the route computation and packet forwarding functions performed by traditional routers into logical entities that perform each function separately.

VP Cross-Connect. An ATM network element that performs VPI translation but does not provide signalling functionality.

VP Switch. An ATM network element that performs VPI translation and provides signalling functionality.

W

Weighted Fair Queuing (WFQ). A packet forwarding technique that allocates bandwidth on an output link among a number of competing output queues.

Weighted Random Early Drop (WRED). A variation of RED that allows different packet drop probabilities to be assigned to each output queue.

X

X.25. An ITU communications standard that defines a data communications protocol between customer equipment and network equipment in a public data network.

INDEX

ARPANET, 278
Adaptive Clock Method, 66, 67, 187
Admission Control, 288—289
Advanced Network Services (ANS), 278
Alarm Indication Signal (AIS), 43, 129
Alignment Stage, 104
Allowable Cell Rate (ACR), 154—155
Answer Supervision, 190
Application Entity (AE), 117
Application Process (AP), 117
Application Service Entity (ASE), 117
Associated Mode Signalling, 120
Association Control Service Element (ACSE), 117
Assured Data Transfer, 101
Asymmetric Digital Subscriber Line (ADSL), 342—343
Asynchronous Transfer Mode (ATM)
  cross-connect, 24—25
  defined, 2
  IP multicast over, 241—244
  TCP performance
    experimental, 266
    impact of switch buffer size on, 268—270, 272—274
    simulation, 267—274
    tuning, 275
  relation to
    B-ISDN, 8
    local area networks, 10
    public network infrastructure, 10
    wide area networks, 10
  switch (network element)
    defined, 24—25
    comparison with router, 314—315
  switching
    virtual channel switching, 21
    virtual path switching, 20—21

Asynchronous Transfer Mode (ATM) (*Cont.*):
  transport structure
    ATM layer 16, 17
    physical layer, 16, 17—18
  trends, 10—11
    bandwidth needs, 11
    QoS, 11
  voice traffic
    interworking to the desktop, 320
    PDU size, 320—321
    signalling interworking, 321—322
ATM Adaptation Layer (AAL)
  AAL 0, 64
  AAL 1, 63, 65—68
  AAL 2, 64, 68—79
  AAL 3/4, 64, 79—82
  AAL 5, 64, 85—88
  defined, 53, 63—65
  parameter negotiation, 109
  sublayers of, 56
ATM Address
  formats, 217—218
  registration, 138
ATM Address Resolution Protocol (ATMARP), 226
ATM Cell
  partial fill, 181
  structure defined, 58—62
ATM Forum, 16
  implementation agreements
    UNI 3.1, 109
    UNI 4.0, 109—110
ATM Layer
  defined, 53, 58—62
  management plane interface, 134—136
  service access point, 62
ATM Interworking
  benefits of, 174
ATM Network Management
  activities, 40—44

ATM Network Management (*Cont.*):
  element management system (EMS), 137
  interfaces
    M1, M2, M3, M4, M5, 39
  layer management, 124—125
  network management system (NMS), 136—137
  plane management, 125—126
  reference model, 37, 38
  relation to the TMN model 38, 39
ATM Service Categories, 150—151
  factors influencing selection of, 151
ATM Network Service Registry, 138
ATM Routing, *see* Private Network to Network Interface
ATM Signalling
  associated, 97
  nonassociated, 97
  endpoints, 33
  interfaces, 92, 93, 95—96
  meta-signalling, 34
  VCC for, 97
  user to network, 34, 35
  user to user, 34, 36
ATM Trunking, 188—193
ATM User to User (AUU), 62
Available Bit Rate (ABR), 151, 152—155
  control loop for, 264

Backward Error Congestion Notification (BECN)
  in frame relay
    defined, 197—198
    mapping to ATM EFCI, 200—201
  in resource management cell, 153
Bandwidth Management, 341—342
Bearer Connection Control, 117
Bearer Services
  defined, 30
Bellcore
  technical reference TR-TSV-000772, 208
Best Effort Service, 282
  interworking with ATM, 305
Broadband High-Layer Information, 193, 332

Broadband Integrated Services Digital Network (B-ISDN)
  defined, 8
  network architecture, 21—29
    network elements, 23—25
    reference points, 24
  objectives, 8
  protocol reference model (PRM), 52—54
  relation to ATM, 8
Broadband ISDN User Part (B-ISUP), 96
  procedures, 116
  specification model, 117—118
Broadband Inter-Carrier Interface (B-ICI)
  defined, 23
Broadband Low-Layer Information, 251, 332
Broadband Message Transfer Part at Level 3 (BMTP-3), 96
Broadband Network Termination 1 (B-NT1)
  defined, 23
Broadband Network Termination 2 (B-NT2)
  defined, 23
Broadband Terminal Adaptor (B-TA)
  defined, 24
Broadband Terminal Equipment Type 1 (B-TE1)
  defined, 23
Broadband Terminal Equipment Type 2 (B-TE2)
  defined, 23
Burst Tolerance, 163

Cache Imposition, 247
Call Control, 117
Cell (*see also* ATM Cell)
  tagging, 162
  relation to cell relay service, 7
  size of, 7
Cell Delay Variation (CDV), 31, 143—144
  measurement of, 146—147
  relation to CDVT, 161—162
  sources of, 146

Index

Cell Delay Variation Tolerance (CDVT), 161—162
Cell Dropping Strategies, 263—266, 271—272
Cell Error Ratio (CER), 143, 144
Cell Loss Ratio (CLR), 31, 143—144
  measurement of, 146
Cell Misinsertion Rate (CMR), 143, 145
  measurement of, 146
Cell Relay Service, 7
Cell Transfer Delay (CTD), 31, 143—144
  sources of, 145
Changeover Procedure, 108
Channel Associated Signalling (CAS), 179—181
Channel Service Unit (CSU), 175
Circuit Emulation Service (CES)
  clocking, 181, 186—187
  delay associated with, 181
  service definition, 194—195
  structured, 177, 179—186
  timeslot mapping, 181, 182, 183
  unstructured, 177, 186—187
  variable bit rate, 187
Circuit Switching, 2—3, 174
  ATM interworking, see Circuit Emulation Service
Classical IP over ATM
  ATMARP server, 227
  logical IP subnet (LIS), 226
  one-armed router, 230
Cluster, 243
Commercial Internet Exchange (CIX), 279
Common Part Convergence Sublayer (CPCS), 64
Conformance Definition, 163
Congestion Control, 142
Connecting Point, 21
Connection Admission Control (CAC), 142, 166—167
Connection Oriented Network Service (CONS), 4
Connectionless Network Access Protocol (CLNAP), 82
Connectionless Network Service (CLNS), 4, 82—84

Connectionless Service Function (CLSF), 82
Constant Bit Rate (CBR) traffic, 31, 150, 151
Continuity Check, 129
Control Loop
  ABR, 153
  TCP, 263
Control Plane (C-plane)
  defined, 54
Controlled Link Sharing, 282
Controlled Load Service, 294
  interworking with ATM, 305
Convergence Sublayer, 56
Cut Through Route, 230

DTMF Signalling, 188
Data Communications, 174
Data Link Connection Identifier (DLCI), 195—196
Data Service Unit (DSU), 175
Defense Advanced Research Projects Agency (DARPA), 256
Dense Wave Division Multiplexing (DWDM), 340—341
Digital Section
  defined, 17—18
  F2 oam flows, 127
  relation to SDH, SONET, 18
Digital Signal Levels (DSn), 5
  DS0, 5, 7
  DS1, 5, 6, 8
  DS3, 5, 6
  TC sublayer functions, 58
Digital Television (DTV), 343—344
Discard Eligibility (DE)
  defined, 198
  mapping to CLP, 198—199
Dynamic Structure Sizing Method, 187

Early Packet Discard (EPD), 265, 266
Edge Device, 176, 244
Ethernet Address, 216
Explicit Forward Congestion Indication (EFCI), 61
Extended Superframe (ESF), 182

F.R.F.5, 201
F.R.F.8, 203—204
Fault Recovery, 44
Federal Internet Exchange (FIX), 279
Flow Control, 76
Flow ID, 288
Forward Explicit Congestion Notification (FECN)
　defined, 196—197
　mapping to EFCI, 199—200
Frame
　relation to frame relay service, 7
Frame Based User to Network Interface (FUNI)
　defined, 206
　frame format, 206—207
Frame Discard, 168
Frame Relay Service, 7, 176
　ATM interworking, 195
　frame structure, 195—198
　service definition, 204—205

Generic Cell Rate Algorithm (GCRA), 157, 158, 159—160
Generic Flow Control (GFC), 59
Goodput, 264
Guaranteed Frame Rate (GFR), 168—169
　relation to FUNI, 207
Guaranteed Service, 293—294
　interworking with ATM, 303—305

H.323
　ATM interworking
　　broadband high-layer information, 332
　　broadband low-layer information, 332
　　fast connect procedure, 333
　　gateway to non-H.323 network, 335
　　RTP media channel, 331
　comparison between ATM and IP, 333—334
　H.225 call signalling channel, 325
　H.245 control channel, 327
　media channel, 328
　network elements, 323—324

H.323 (Cont.):
　registration, admission, and status channel (RAS), 324
　terminal capabilities exchange, 327—328
　zone, 324
Header
　in ATM cell structure, 59
Header Error Correction (HEC)
　in ATM cell structure, 62
Heterogeneous Branch Point, 291
High Bit Rate Digitial Subscriber Line (HDSL), 342
High Definition Television (HDTV), 343
Hub Site, 175

IBM Corporation, 278
IEEE 802.1a, 250
IEEE 802.2, 250
Idle Cell, 59, 60
Information Distribution Services, 9—10
Integrated Internet Services (IIS) model
　interworking with ATM, 297—298
　IP flows, 298—299
　heterogeneous branch points, 299—301
　RSVP signalling, 302—303
　RSVP state, 301—302, 303
　traffic control functions, 282
Integrated Link Management Interface (ILMI), 137—138
Integrated Services Digital Network (ISDN), 8, 175
　basic rate interface (BRI), 195
　digital subscriber line (DSL), 342
　primary rate interface (PRI), 195
　signalling, 191—193
Interactive Services
　distance learning, 9
　Message services, 9
　videoconferencing, 8
International Telecommunication Union (ITU), 16
　recommendations
　　G.711, 320

Index

International Telecommunication Union (ITU), recommendations (*Cont.*):
  H.225, 325—326
  H.245, 327—328
  H.323, 322—324
  I.321, 52
  I.361, 58
  I.363, 63
  I.363.2, 63
  I.364, 82
  I.366.1, 63
  I.432, 56
  Q704, 103
  Q931, 191
  Q1400, 117
  Q2110, 72
  Q2764, 120
  Q2931, 95
  Q2971, 109
Internet, the
  architecture, 278—280
  number of networks, 278
  origins, 278
  traffic patterns, 280
Internet Architecture Board (IAB), 280
Internet Engineering Task Force (IETF), 280
  RFC 1112, 241
  RFC 1122, 227
  RFC 1323, 258
  RFC 1483, 250
  RFC 1700, 251
  RFC 2225, 226
Internet Group Management Protocol (IGMP), 241
Internet Protocol (IP), 216
  flow, 282
  SONET versus ATM, 311, 314—315
  protocol efficiency, 313
  maximum transfer unit (MTU), 258
  multicast, 241
IP version 6 (IPv6), 288
Internet Service Provider (ISP), 279
Internet Society, 280
Interworking Function (IWF), 177

LAN Emulation (LANE)
  broadcast and unknown server, 236
  comparison with legacy LAN, 240
  connections used with, 237—237
  LAN emulation client (LEC), 236
  LAN emulation configuration server (LECS), 236
  LAN emulation server (LES), 236
  multiprotocol support, 235
  proxy LEC, 236
  supported LAN topologies, 236
LLC/SNAP Encapsulation, 230
Latency, 7
Leaf-Initiated Join (LIJ)
  root-prompted join, 115
  leaf-prompted join without root notification, 114
Leaf-Node, 29, 112
Line, 18 (same as Digital Section)
Line Terminating Equipment (LTE), 18 (*see also* Digital Section)
Local Multipoint Distribution Service (LMDS), 345—346
Loop, 5
Loop Timing, 181
Loopback, 129—130, 131, 132—133

MCI Incorporated, 278
MPEG-2, 343
Maintenance Control, 117
Managed Object, 39
Management Information Base (MIB), 40
Management Plane (M-plane)
  ATM layer interface, 134—136
  defined, 54
  layer management, 54
  plane management, 54
Maximum Burst Size (MBS), 31, 149
Message Mode Service, 81
Meta-Signalling, 34, 60, 61
Merit Networks, 278
Minimum Cell Rate (MCR), 149
Multicast Address Resolution Server (MARS), 241
  connections, 243—244
Multicast Server (MCS), 241

Index

Multiple Access Protocol over SONET/SDH (MAPOS), 311
Multiplex Section, 18 (same as Digital Section)
Multipoint Connection Point (MPCP), 28—29
Multiprotocol over ATM (MPOA)
  address resolution, 246
  cache imposition, 247
  control flows, 246
  cut through route, 247—248
  data flows, 246
  MPOA client (MPC), 244
    roles, 246
  MPOA server (MPS), 244
  multiprotocol support, 244
  relation to
    LANE, 244
    NHRP, 244
Multiprotocol Device, 244
Multiprotocol Encapsulation
  LLC/SNAP, 250, 251
  VC based multiplexing, 250, 251

National Science Foundation (NSF), 278
NSFNET, 278
Network Access Point (NAP), 279
  ATM deployment at, 308—310
Network Driver Interface Specification (NDIS), 235
Network Leaf-Initiated Join, 114
Network Node Interface (NNI)
  ATM cell structure, 59
  defined, 23
Network Parameter Control (NPC), 33, 157
  location of, 156
Network Resource Management, 142
Network Service Access Point (NSAP) address, 216
Network Service Provider (NSP), 279
Next Hop Resolution Protocol (NHRP)
  address resolution, 233
  compared to Classical IP over ATM, 230

Next Hop Resolution Protocol (NHRP) (*Cont.*):
  connectionless data transport, 234
  next hop client (NHC), 231
  next hop server (NHS), 231
  relation to OSPF, 252
Nodal Functions, 117
Non-Broadcast Multiple Access (NBMA) network, 230
Nonconforming Traffic, 140
Null AAL, 64

One Armed Router, 230
Open Data-Link Interface (ODI), 235
Operations Administration, Maintenance, and Provisioning (OAM&P) (*see also* ATM Network Management)
  configuration management, 136
  defined 36, 126
  functions
    ATM layer, 129—130
  hierarchical structure, 127
  loopback, 129—130, 131, 132—133
  performance monitoring, 130, 132, 133—134
OAM Flows
  ATM layer, 128—129
  cell-based, 128
  defined, 126—127
  end-to-end, 129
  physical layer, 128
  SDH, 128
  segment, 129
  SONET, 128
OAM Cell
  defined, 42
  formats, 130—134
  types, 43, 44, 60, 129—130
Open Shortest Path First (OSPF), 233, 252
Open Systems Interconnection reference model (OSIRM)
  application layer, 52
  data link layer, 52
  network layer, 52
  physical layer, 52
  presentation layer, 52

Index

Open Systems Interconnection
 reference model (OSIRM) (*Cont.*):
 session layer, 52
 transport layer, 52
Packet Classification, 288
Packet Discard, 285
Packet Forwarding, 244
 delay, 285
Packet Scheduler, 283
Packet Switching, 3—5
Partial Packet Discard (PPD), 168,
 264—265, 266
Path, 17 (same as Transmission Path)
Path Terminating Equipment (PTE),
 17 (*see also* Transmission Path)
Payload
 in ATM cell structure, 59
Payload Type Identifier (PTI), 61
Peak Cell Rate (PCR), 31, 149
Peer Group Leader (PGL), 219
Peering, 279
Performance Monitoring, 130, 132,
 133—134
Permanent Virtual Connection
 (PVC), 20
Personal Communication Services
 (PCS), 345
Physical Layer, 56—58
 relation to B-ISDN PRM, 53
 SAP defined, 58
 sublayers, 57—58
Physical Media Dependent (PMD)
 sublayer
 defined, 57
 relation to physical layer, 56—57
Plesiochronous Digital Hierarchy
 (PDH), 5—6, 17
Point to Point Protocol (PPP), 311, 312
Primitives, 56
Private Networks, 175
Private Network to Network
 Interface (PNNI)
 ATM routing
  attributes, 226
  comparison with OSPF, 252
  designated transit list (DTL), 224
  flooding, 221—222
  functions, 220

Private Network-to-Network Interface
 (PNNI), ATM routing (*Cont.*):
 hello protocol, 220, 222
 hierarchy, 219
 metrics, 225
 relation to QoS, 226
 ATM signalling, 115—116
Protocol Data Unit (PDU)
 at the AAL-SAP, 56
Public Network Service, 175—176

Quality of Service (QoS)
 factors impacting, 147—148
 parameters, 31, 143
 reference interface for
  measurement of, 32
 relation to traffic contract, 148
Quasi-Associated Mode Signalling, 120

Random Early Drop (RED), 286
Real Time Control Protocol (RTCP),
 329
Real Time Protocol (RTP), 328—329
Regenerator Section
 defined 18,
 F1 oam flows, 127
 relation to
  LTE, 18
  PTE, 18
  SONET, 18
  STE, 18
Reliable Transport, 325
Remote Defect Indication (RDI),
 44, 129
Reshaping Point, 291
Resource Management Cell, 60
 format, 153—155
 use with ABR service, 151, 152
Resource Reservation Protocol
 (RSVP)
 adspec, 293—294
 comparison to ATM, 290—291
 filter spec, 290
 flowspec, 295—297
 overview, 289—290
 path message, 290, 291—292
 receiver, 290
 rcsv message, 290, 295

Resource Reservation Protocol (RSVP) (*Cont.*):
  sender, 290
  session creation, 297
  tspec, 292—293
Robbed Bit Signalling, 179, 188
Root-Initiated Join, 112
Root Leaf-Initiated Join, 115
Root Node, 29, 112
Route Computation, 244
Router, 244
  compared with ATM switch, 314—315
Routing Arbiter (RA), 280

Section, 18 (same as Regenerator Section)
Section Terminating Equipment (STE), 18 (*see also* Regenerator Section)
Segmentation and Reassembly (SAR), 56, 64
Service Access Point (SAP)
  defined at the AAL (AAL-SAP), 56
Service Data Unit (SDU)
  at the AAL-SAP, 56
Service Level Agreement (SLA), 342
Service Specific Assured Data Transfer Sublayer (SSADT), 72
Service Specific Connection Oriented Protocol (SSCOP)
  example, 76—78
  functions performed by, 98—102
  PDU types, 72—76
Service Specific Convergence Function (SSCF)
  NNI, 94, 103—108
  UNI, 94, 102—103
Service Specific Convergence Sublayer (SSCS), 64
Service Specific Segmentation and Reassembly Sublayer (SSSAR), 72
Service Specific Transmission Error Detection Sublayer (SSTED), 72
Severely Errored Cell Block Ratio (SECBR), 143, 145
  measurement of, 146
Simple and Efficient AAL (SEAL), 85

Signalling ATM Adaptation Layer (SAAL)
  defined, 92, 94
Signalling System Number 7 (SS7), 120
Single Association Control Function (SACF), 117
Single Association Object (SAO), 117
Sliding Window, 76
Soft PVC, 137
Streaming Mode Service, 81
Structured Data Transfer
  defined in AAL 1, 65, 67
Subnetwork Attachment Point (SNAP) address, 217
Supervisory Signalling, 188—190
Sustained Cell Rate (SCR), 31, 149
Swedish University Network (SUNET), 266
Switched Multimegabit Data Service (SMDS), 176, 207—208
  ATM interworking, 210—211
  service definition, 211—212
Switched Virtual Connection (SVC)
  defined, 20
Synchronization
  of packet flows, 286
Synchronous to Asynchronous Conversion (SAC), 189
Synchronous Digital Hierarchy (SDH), 6, 17, 18
Synchronous Optical Network (SONET), 6, 17, 18
  hierarchy, 6
  Synchronous Payload Envelope (SPE), 311
  TC sublayer functions, 58
  transmission rates, 6
Synchronous Residual Time Stamp (SRTS), 66
  in circuit emulation service, 186—187
  use in measurement of CDV, 147
Synchronous Transfer Signal (STS), 6

Tail Drop, 286
Telecommunications Management Network (TMN)
  interfaces
    Q_3, 39

Index

Telecommunications Management
 Network (TMN),
 interfaces (*Cont.*):
  X, 39
  model, 37—38
Teleservices
 defined, 30
Time Division Multiplexing (TDM), 2
Traffic Conformance
 enforcement mechanisms, 32—33
 generic cell rate algorithm, 157
 relation to QoS, 32
Traffic Contract, 31, 148
Traffic Descriptor, 148
Traffic Parameters, 148—149
Traffic Shaping, 142, 167—168
Transfer Mode, 2
Translational Bridging, 239
Transmission Control Protocol
  (TCP), *see also* Asynchronous
  Transfer Mode
 acknowledgement, 256
 bandwidth-delay product, 258, 259
 congestion avoidance, 260—261
 congestion control, 257
 control loop, 263
 fast retransmit, 261—262
 flow control, 256
 interaction with ATM, 256
 maximum segment size (MSS),
  258—259
 network congestion, 257, 341
 over native ATM, 274—275
 packet loss, 260—261
 performance, 262
 round trip time (RTT), 257
 sliding window, 256
 slow start, 260
 Tahoe version, 260
 timeout, 257, 261—262
 window size, 257—258
Transmission Convergence (CS)
  sublayer
 defined, 57—58
 relation to physical layer, 56—57
Transmission Path
 defined, 17
 F3 oam flows, 127

Transmission Path (*Cont.*):
 relation to PDH, SDH, SONET, 17
Transmission Systems
 defined, 5
 digital, 5
 hierarchies, 5—6
 T-carrier, 5
Transport Address, 324
Trunk, 5

Unreliable Transport, 324
Unspecified Bit Rate (UBR), 151
Unstructured Data Transfer
 defined in AAL 1, 65
Usage Parameter Control (UPC), 32,
  33, 155—157
 location of, 156
User Datagram Protocol (UDP),
  266, 341
User Network Interface (UNI)
 ATM cell structure, 59
 defined, 23
 point-to-multipoint connection,
  28—30
  leaf-initiated join, 114—115
  root-initiated join, 112—114
 point-to-point connection, 27—28
  connection clearing, 111—112
  connection establishment,
   110—111
 signalling
  capabilities, 109
User Plane (U-plane)
 defined, 53—54

Variable Bit Rate (VBR) traffic, 31
 non-real time (nrt-VBR), 150—151
 real time (rt-VBR), 150, 151
very high performance Backbone
  Network Service (vBNS), 279
 bandwidth management, 308
 deployment of ATM, 306—308
Video Distribution, 344
Virtual Channel (VC)
 defined, 17
VC Based Multiplexing, 250, 251
Virtual Channel Connection (VCC)
 defined, 18 19

Virtual Channel Connection (VCC) (*Cont.*):
  endpoint, 19
  F5 oam flows, 127
Virtual Channel Identifier (VCI)
  defined, 17
  field in ATM cell structure, 59
  reserved values, 60
Virtual Channel Link (VCL), 21
Virtual LAN, 230
Virtual Path (VP)
  defined, 17
Virtual Path Connection (VPC)
  defined, 18
  endpoint, 18
  F4 oam flows, 127
Virtual Path Connection Identifier (VPCI)
  defined, 21
  relation to
    VPC, 21

Virtual Path Connection Identifier (VPCI), relation to (*Cont.*):
  switching, 21
  network management, 21
Virtual Path Identifier (VPI)
  defined, 17
  field in ATM cell structure, 59
  reserved values, 60
Virtual Path Link (VPL), 21
Virtual Routing, 244
Virtual Scheduling Algorithm, 157—159
Virtual Studio, 344
Weighted Fair Queuing (WFQ), 283—285
Weighted Random Early Drop (WRED), 287
Wireless ATM, 344—346
World Wide Web (WWW), 280

X.25, 7, 174—175

ABOUT THE AUTHOR

Ronald H. Davis is a member of the technical staff at Lucent Technologies. His experience includes network architecture, broadband switching technologies, and ATM switch software development. Previous experience includes product marketing in the semiconductor industry at NCR Microelectronics, Advanced Micro Devices, and Intel. He holds a master's degree in electrical engineering from the Massachusetts Institute of Technology and an MBA from the Stanford Business School.

DATE DE RETOUR		L.-Brault
2 4 NOV. 2005		